William Miller
Rita M. Pellen
Editors

Dealing with Natural Disasters in Libraries

Dealing with Natural Disasters in Libraries has been co-published simultaneously as *Public Library Quarterly*, Volume 25, Numbers 3/4 2006.

Pre-publication
REVIEWS,
COMMENTARIES,
EVALUATIONS . . .

"A VALUABLE READ FOR ANYONE WHO WORKS IN ANY LIBRARY, regardless of type, size, or budget. The collection's strength is that it goes beyond useful, general advice to offer real-life lessons and stories from information professionals who have been on the front lines during disasters, and helped their institutions and communities recover from the worst."

Danielle Pollock, MA
Librarian
Missouri Department of Transportation

More pre-publication
REVIEWS, COMMENTARIES, EVALUATIONS . . .

"Hurricane Katrina was a wakeup call for libraries all over North America. Are we prepared for a disaster of that magnitude? But it is just as important to prepare ourselves for the smaller disasters, the fires, the broken pipes, the mold, which are more likely to affect our libraries. No library is immune to natural disasters, whether the threat comes in the form of a fire that destroys an office and damages part of the collection, a mold growth that affects an academic library branch, or a hurricane that devastates a region and destoys most of its libraries. *Dealing with Natural Disasters in Libraries* collects the stories of library workers from a variety of libraries affected by disaster. The stories are a mix of personal anecdotes, lessons learned, and advice for other library workers. Some focus on describing their activities in the first month after a disaster, while others focus on describing the lessons they learned from dealing with a disaster. If the disaster affected more than the library, some articles include useful advice on how to tactfully protect the library from quick political decisions that might have unforeseen consequences. The chapters in this book also describe the emotional effects of dealing with disasters, and some elements that are sometimes not taken into account in disaster planning. For example, library staff might be dealing with property loss or injuries or worse from a disaster. The staff might not even be allowed in the building. In addition, the writers analyze events and decisions that affected their ability to manage the disaster. Details like this make *Dealing with Natural Disasters in Libraries* a valuable addition to personal and professional collections."

Jennifer Crispin, MAISLT (MLS)
Doctoral Student in Information Science and Learning Technologies, University of Missouri-Columbia

More pre-publication
REVIEWS, COMMENTARIES, EVALUATIONS . . .

"GIVES PRACTICAL ADVICE AND STRATEGIES for recovering successfully from disasters of all types, and more importantly, mitigating damage through careful planning. While the profiles of recovered libraries are a unique resource, Paul T. Jaeger et al.'s paper on the evolving roles of libraries in hurricane recovery demonstrates an even more compelling portrait of the library as a community recovery center that connects people to family members, government information, and recovery resources. . . . AN IMPORTANT ADDITION TO THE LIBRARY LITERATURE and will help libraries map out our role within the community in the coming years."

Carrie Lowe, MLS,
Internet Policy Specialist,
American Library Association Office for Information Technology Policy

"AN IMPORTANT ADDITION TO THE LIBRARY LITERATURE. . . . From the Pacific Islands to the Northeast and throughout the Gulf Coast, they remind us again of how important preparedness and planning, establishing collaborative relationships, and staff training are to a successful recovery. In these pages are the words of wisdom from those who are well qualified to support the critical aspects of what to do *before* disaster strikes. THE INFORMATIVE AND WELL-WRITTEN CASE STUDIES GIVE GOOD ADVICE AND CONCRETE EXAMPLES BASED ON RECENT EVENTS; hopefully these will motivate libraries to take disaster preparedness actions! The bibliography provides a helpful combination of more disaster stories and planning resources."

Julie A. Page,
Co-Coordinator,
California Preservation Program, California Alliance for Response, and Western States & Territories Preservation Assistance ServiceText

Dealing with Natural Disasters in Libraries

Dealing with Natural Disasters in Libraries has been co-published simultaneously as *Public Library Quarterly*, Volume 25, Numbers 3/4 2006.

Monographic Separates from *Public Library Quarterly*®

For additional information on these and other Haworth Press titles, including descriptions, tables of contents, reviews, and prices, use the QuickSearch catalog at http://www.HaworthPress.com.

Dealing with Natural Disasters in Libraries, edited by William Miller and Rita M. Pellen (Vol. 25, No. 3/4, 2006). *A comprehensive examination of case studies detailing how libraries have responded to a variety of natural disasters, with suggestions for future planning and response.*

Current Practices in Public Libraries, edited by William Miller and Rita M. Pellen (Vol. 25, No. 1/2, 2006). *Examines a variety of current trends, issues, and practices in public library administration.*

Dealing with Natural Disasters in Libraries

William Miller
Rita M. Pellen
Editors

Dealing with Natural Disasters in Libraries has been co-published simultaneously as *Public Library Quarterly*, Volume 25, Numbers 3/4 2006.

The Haworth Information Press®
An Imprint of The Haworth Press, Inc.

www.HaworthPress.com

Published by

The Haworth Information Press®, 10 Alice Street, Binghamton, NY 13904-1580 USA

The Haworth Information Press® is an imprint of The Haworth Press, Inc., 10 Alice Street, Binghamton, NY 13904-1580 USA.

Dealing with Natural Disasters in Libraries has been co-published simultaneously as *Public Library Quarterly*, Volume 25, Numbers 3/4 2006.

Cover design by Marylouise Doyle

Cover photos "University of Georgia Library Varieties of Heat Damage" are courtesy of UGA Libraries, Katie Gentilello. 2003. Reprinted with permission.

Cover photo "University of Hawaii Manoa Libratry The Map Collection Room Immediately After the Disaster" is courtesy of Susan Murata and Julie Page. Reprinted with permission.

Cover photo "University of Iowa Books in University Archives with Mold" is ©University of Iowa Libraries. Reprinted with permission.

Library of Congress Cataloging-in-Publication Data

Dealing with natural disasters in libraries / William Miller, Rita M. Pellen, editors.
p. cm.
"Co-published simultaneously as Public library quarterly, volume 25, numbers 3/4 2006."
Includes bibliographical references and index.
ISBN-13: 978-0-7890-3609-4 (alk. paper)
ISBN-13: 978-0-7890-3610-0 (pbk. : alk. paper)
1. Libraries–Safety measures–Planning. 2. Library materials–Conservation and restoration–Planning. 3. Emergency management–United States–Case studies. 4. Disaster relief–United States–Case studies. I. Miller, William, 1947- II. Pellen, Rita M. III. Public library quarterly (New York, N.Y.)
Z679.7.D43 2007
025.8'2–dc22

2007033831

The HAWORTH PRESS *Inc*

Abstracting, Indexing & Outward Linking

PRINT *and* ELECTRONIC BOOKS & JOURNALS

This section provides you with a list of major indexing & abstracting services and other tools for bibliographic access. That is to say, each service began covering this periodical during the year noted in the right column. Most Websites which are listed below have indicated that they will either post, disseminate, compile, archive, cite or alert their own Website users with research-based content from this work. (This list is as current as the copyright date of this publication.)

Abstracting, Website/Indexing Coverage . . . . Year When Coverage Began

- ***International Bibliography of Book Reviews on the Humanities and Social Sciences (IBR) (Thomson) <http://www.saur.de>*. . . 2006**
- *****Academic Search Premier (EBSCO)** <http://search.ebscohost.com>*. 2006**
- *****LISA: Library and Information Science Abstracts (ProQuest CSA)** <http://www.csa.com/factsheets/list-set-c.php>*. . 1992**
- *****MasterFILE Premier (EBSCO)** <http://search.ebscohost.com>*. 2006**
- ***Academic Source Premier (EBSCO) <http://search.ebscohost.com>*. 2007**
- ***Advanced Polymers Abstracts (ProQuest CSA) <http://www.csa.com/factsheets/ema-polymers-set-c.php>* 2006**
- ***Aluminium Industry Abstracts (ProQuest CSA) <http://www.csa.com/factsheets/aia-set-c.php>* 2006**
- ***British Library Inside (The British Library) <http://www.bl.uk/services/current/inside.html>*. 2006**
- ***Cabell's Directory of Publishing Opportunities in Educational Technology & Library Science <http://www.cabells.com>* 2006**
- ***Cambridge Scientific Abstracts (now ProQuest CSA) <http://www.csa.com>* 2006**

(continued)

- *Ceramic Abstracts (ProQuest CSA) <http://www.csa.com/factsheets/wca-set-c.php>* **2006**
- *Composites Industry Abstracts (ProQuest CSA) <http://www.csa.com/factsheets/ema-composites-set-c.php>* . . . **2006**
- *Computer and Information Systems Abstracts (ProQuest CSA) <http://www.csa.com/factsheets/computer-set-c.php>* **2004**
- *Corrosion Abstracts (ProQuest CSA) <http://www.csa.com/factsheets/corrosion-set-c.php>* **2006**
- *CSA Engineering Research Database (ProQuest CSA) <http://www.csa.com/factsheets/engineering-set-c.php>* **2006**
- *CSA High Technology Research Database With Aerospace (ProQuest CSA) <http://www.csa.com/factsheets/hightech-set-c.php>* **2006**
- *CSA Technology Research Database (ProQuest CSA) <http://www.csa.com/factsheets/techresearch-set-c.php>* **2006**
- *CSA/ASCE Civil Engineering Abstracts (ProQuest CSA) <http://www.csa.com/factsheets/civil-set-c.php>* **2006**
- *Current Abstracts (EBSCO) <http://search.ebscohost.com>* **2007**
- *Current Citations Express (EBSCO) <http://search.ebscohost.com>* **2007**
- *EBSCOhost Electronic Journals Service (EJS) <http://search.ebscohost.com>* **2001**
- *Electronic Collections Online (OCLC) <http://www.oclc.org/electroniccollections/>* **2006**
- *Electronics and Communications Abstracts (ProQuest CSA) <http://www.csa.com/factsheets/electronics-set-c.php>* **2006**
- *Elsevier Eflow-D <http://www.elsevier.com>* **2006**
- *Elsevier Scopus <http://www.info.scopus.com>* **2005**
- *Engineered Materials Abstracts (ProQuest CSA) <http://www.csa.com/factsheets/emaclust-set-c.php>* **2006**
- *Google <http://www.google.com>* **2004**
- *Google Scholar <http://scholar.google.com>* **2004**
- *Haworth Document Delivery Center <http://www.HaworthPress.com/journals/dds.asp>* **1979**

(continued)

- *Index Guide to College Journals* **1999**
- *Index to Periodical Articles Related to Law <http://www.law.utexas.edu>* **1992**
- *Informed Librarian, The <http://www.informedlibrarian.com>* . . . **1993**
- *INIST-CNRS <http://www.inist.fr>* **2000**
- *International Bibliography of Periodical Literature on the Humanities and Social Sciences (Thomson) (IBZ) <http://www.saur.de>* **1996**
- *Internationale Bibliographie der geistes- und sozialwissenschaftlichen Zeitschriftenliteratur . . . See IBZ <http://www.saur.de>* **2001**
- *Journal of Academic Librarianship: Guide to Professional Literature, The* **2006**
- *JournalSeek <http://www.journalseek.net>* **2006**
- *Konyvtari Figyelo (Library Review) <http://www.oszk.hu/index_en.html>* **1997**
- *Library Literature & Information Science Index / Full Text (H.W. Wilson) <http://www.hwwilson.com>* **1984**
- *Library, Information Science & Technology Abstracts (EBSCO) <http://search.ebscohost.com>* **2006**
- *Library, Information Science & Technology Abstracts with Full Text (EBSCO) <http://search.ebscohost.com>* **2007**
- *Links@Ovid (via CrossRef targeted DOI links) <http://www.ovid.com>* **2005**
- *Materials Business File (ProQuest CSA) <http://www.csa.com/factsheets/mbf-set-c.php>* **2006**
- *Materials Research Database with METADEX (ProQuest CSA) <http://www.csa.com/factsheets/materials-set-c.php>* **2006**
- *Mechanical & Transportation Engineering Abstracts (ProQuest CSA) <http://www.csa.com/factsheets/mechtrans-set-c.php>* **2006**
- *METADEX (ProQuest CSA) <http://www.csa.com/factsheets/metadex-set-c.php>* **2006**
- *NewJour (Electronic Journals & Newsletters) <http://gort.ucsd.edu/newjour/>* **2006**
- *OCLC ArticleFirst <http://www.oclc.org/services/databases/>* **2007**

(continued)

- *Ovid Linksolver (OpenURL link resolver via CrossRef targeted DOI links) <http://www.linksolver.com>* **2005**
- *Scopus (See instead Elsevier Scopus) <http://www.info.scopus.com>* **2005**
- *Solid State and Superconductivity Abstracts (ProQuest CSA) <http://www.csa.com/factsheets/solid-state-set-c.php>* **2006**
- *SwetsWise <http://www.swets.com>* **2000**
- *TOC Premier (EBSCO) <http://search.ebscohost.com>* **2007**
- *WilsonWeb <http://vnweb.hwwilsonweb.com/hww/Journals/>* **2005**
- *zetoc (The British Library) <http://www.bl.uk>* **2004**

Bibliographic Access

- ***Cabell's Directory of Publishing Opportunities in Educational Curriculum and Methods <http://www.cabells.com/>***
- ***MediaFinder <http://www.mediafinder.com/>***
- ***Ulrich's Periodicals Directory: The Global Source for Periodicals Information Since 1932 <http://www.bowkerlink.com>***

Special Bibliographic Notes related to special journal issues (separates) and indexing/abstracting:

- indexing/abstracting services in this list will also cover material in any "separate" that is co-published simultaneously with Haworth's special thematic journal issue or DocuSerial. Indexing/abstracting usually covers material at the article/chapter level.
- monographic co-editions are intended for either non-subscribers or libraries which intend to purchase a second copy for their circulating collections.
- monographic co-editions are reported to all jobbers/wholesalers/approval plans. The source journal is listed as the "series" to assist the prevention of duplicate purchasing in the same manner utilized for books-in-series.
- to facilitate user/access services all indexing/abstracting services are encouraged to utilize the co-indexing entry note indicated at the bottom of the first page of each article/chapter/contribution.
- this is intended to assist a library user of any reference tool (whether print, electronic, online, or CD-ROM) to locate the monographic version if the library has purchased this version but not a subscription to the source journal.
- individual articles/chapters in any Haworth publication are also available through the Haworth Document Delivery Service (HDDS).

As part of Haworth's continuing commitment to better serve our library patrons, we are proud to be working with the following electronic services:

AGGREGATOR SERVICES

- EBSCOhost
- Ingenta
- J-Gate
- Minerva
- OCLC FirstSearch
- Oxmill
- SwetsWise

LINK RESOLVER SERVICES

- 1Cate (Openly Informatics)
- ChemPort (American Chemical Society)
- CrossRef
- Gold Rush (Coalliance)
- LinkOut (PubMed)
- LINKplus (Atypon)
- LinkSolver (Ovid)
- LinkSource with A-to-Z (EBSCO)
- Resource Linker (Ulrich)
- SerialsSolutions (ProQuest)
- SFX (Ex Libris)
- Sirsi Resolver (SirsiDynix)
- Tour (TDnet)
- Vlink (Extensity, formerly Geac)
- WebBridge (Innovative Interfaces)

ALL HAWORTH INFORMATION PRESS BOOKS AND JOURNALS ARE PRINTED ON CERTIFIED ACID-FREE PAPER

Dealing with Natural Disasters in Libraries

CONTENTS

ABOUT THE EDITORS

William Miller, PhD, MLS, is Director of Libraries at Florida Atlantic University. He formerly served as Head of Reference at Michigan State University, and as Associate Dean of Libraries at Bowling Green State University (Ohio). Dr. Miller has a PhD from the University of Rochester and an MLS from the University of Toronto. He is Past-President of the Association of College and Research Libraries, has served as Chair of the *Choice* Editorial Board, and is a frequent contributor to professional journals, as well as being a contributing editor of *Library Issues*. Dr. Miller teaches courses in English Literature and Library Science. He was named Instruction Librarian of the Year in 2004 by the ACRL Instruction Section.

Rita M. Pellen, MLS, is Associate Director of Libraries at Florida Atlantic University. Previously she was Assistant Director of Public Services and Head of the Reference Department at Florida Atlantic. Ms. Pellen has a BA from Pennsylvania State University and an MLS from the University of Pittsburgh. In 1993, she received the Gabor Exemplary Employee Award in recognition for outstanding service to FAU, and in 1997, the "Literati Club Award for Excellence" for the outstanding paper in The Bottom Line. She has served on committees in LAMA, ACRL and ALCTS, as well as the Southeast Florida Library Information Network, SEFLIN, a multi-type library cooperative in South Florida. Honor society memberships include Beta Phi Mu and Phi Kappa Phi.

Introduction: Dealing with Natural Disasters

Most areas of the world will eventually suffer from some major disaster, such as a hit by a comet or asteroid, the effects of global warming or a new ice age, volcanic activity, or a super-tsunami. Beyond that, there are the truly mega-disasters like an encounter with a black hole, a gamma ray shower, or the ultimate demise of the sun, which would end life on the planet and which no one can do anything about (short of relocating to another planet, which itself will not ultimately help, given what physicists tell us about the ultimate dissolution of the universe). Most of these disasters will not happen in our lifetimes, however, and are chiefly the stuff of amusement for those who like watching disaster shows on TV. We are much more likely to encounter a wide range of limited natural disasters, such as flood and fire, which are more local and comparatively less catastrophic (though anyone caught in such a disaster would find it more than sufficiently destructive), and which librarians really can plan for and do something about; it is these kinds of disasters that are the focus of this collection.

Natural disasters are no respecters of type-of-library; they are equal-opportunity destroyers and can occur in any kind of library at any time, and the methods used to deal with them are largely the same, regardless of which kind of library one works in; hence, all types of libraries are discussed in this collection. Natural disasters are especially problematic for cultural institutions like libraries, institutions that must attempt to preserve an extensive collection of artifacts, while at the same time trying

---

[Haworth co-indexing entry note]: "Introduction: Dealing with Natural Disasters." Miller, William. Co-published simultaneously in *Public Library Quarterly* (The Haworth Information Press, an imprint of The Haworth Press, Inc.) Vol. 25, No. 3/4, 2006, pp. 1-5; and: *Dealing with Natural Disasters in Libraries* (ed: William Miller and Rita M. Pellen) The Haworth Information Press, an imprint of The Haworth Press, Inc., 2006, pp. 1-5. Single or multiple copies of this article are available for a fee from The Haworth Document Delivery Service [1-800-HAWORTH, 9:00 a.m. - 5:00 p.m. (EST). E-mail address: docdelivery@haworthpress.com].

Available online at http://plq.haworthpress.com

doi:10.1300/J118v25n03_01

also to restore the services they offer, be these electronic or in person. In this regard, libraries differ from most other kinds of institutions. Among the things that librarians will quickly learn in this collection is that because of the need to preserve artifacts, speed in dealing with a disaster is essential, and whatever the normal bureaucratic regime in which the library works, when a natural disaster occurs, all the normal forms, procedures, work order requests, etc., will have to be set aside in order to respond effectively.

Jane Strudwick's bibliography, "A Selected Bibliography of Library Disaster Stories: Before, During, and After," with its metaphor of "stories," encapsulates the contents of this volume, in that stories are what we have here, stories which show us how cultural institutions, primarily libraries, have responded to natural disasters world-wide in recent years. Librarians would do well to go beyond the limits of the case studies in this volume to read some of the items cited by Strudwick, and learn more about what has occurred in Florence, Italy, Aceh, Indonesia, and other areas of the world, and Strudwick usefully cites a wide range of articles and web sites covering all sorts of disaster. Her bibliography cites many "stories" which, while they do not constitute planning, may nevertheless help us "trudge up the learning curve." She also cites disaster manuals and web sites that can help us plan.

Several pieces here are of a general nature. Susan Carol Curzon's "Coming Back from Major Disaster: Month One" discusses the generic steps one should take during the first month of a disaster. Curzon, whose experience includes handling the aftermath of an earthquake that caused major structural damage to her library, feels that the steps one takes in the first month lay the groundwork for the recovery and set the course for future restoration efforts. Randy Silverman, in "The Seven Deadly Sins of Disaster Recovery," bases his article on the work of Peter Waters, book conservator and the recovery coordinator at the Biblioteca Nazionale in Florence after its devastating flood of 1966. Silverman emphasizes human failings, such as "Sloth," which undermines "the degree of human reaction and responsiveness under stressful conditions." Sonya L. Green and Thomas H. Teper, in "The Importance of Disaster Planning for the Small Public Library," stress the things that a small library can do to affect the efficiency of a recovery, and discuss a 2005 state-wide survey of preservation readiness among cultural heritage institutions in Illinois.

Several articles in this collection discuss fire damage, an especially devastating event, even where immediate damage to materials is contained, because smoke damage throughout the entire building can be pervasive and extremely difficult to eradicate. In "Fire at the University of New

Mexico Library," Terry Gugliotta offers a chronicle of a fire that destroyed 30,000 volumes, and discusses how an earlier flood had prepared the library to better handle this new disaster. Ryan Perry, in "The Great University of Georgia Libraries Fire of '03: Lessons Learned and Questions Raised," discusses a multi-million dollar disaster and the ongoing efforts to ameliorate the damage. Unlike most fires, the one at Georgia went out without the use of water, and so the problems discussed here are very different from most fire-related recovery discussions.

Most fire damage, however, does end up causing water damage, and water is a major enemy of libraries, both in and of itself, and because of the mold it can produce. Tara D. Kennedy (Yale University), writing about a "Steamy Situation: Water Emergency in Sterling Memorial Library," discusses an accidental discharge of sprinklers which could have resulted in major water damage, but was controlled and minimized as a result of prompt attention. In "Riding the Surf: Responding to Library Disasters in Island Communities," Lynn Ann Davis uses a surfing metaphor to discuss how her library dealt with a flash flood in 2006, and to discuss how libraries in sparsely-populated areas can "develop a sustainable model . . . for disaster planning and preservation."

It does not take a flood or a burst pipe or accidental sprinkler discharge to cause moisture-related problems, of course. Mold can occur in almost any library, when moisture conditions are right, or rather wrong. In "The Complicated Symptoms of Sick Buildings," Richard P. Widdicombe chronicles a long-running and still ongoing saga of mold caused by poor building design, its interplay with human frailties, both physical and mental, and the difficulties of retrofitting an inherently flawed facility. Nancy E. Kraft (University of Iowa), in "Major and Minor Mold Outbreaks: Summer of 2002" describes a number of mold outbreaks, both large and small, and discusses the steps used at her library for successful mold treatment. Cynthia Dudenhöffer, in "Keeping the Beast at Bay: Fighting Mold at the University of Missouri-Columbia Journalism Library" discusses a successful three-week effort to control and eliminate an outbreak.

The most well-known and still ongoing cause of mold and water damage, along with structural damage, in recent American history was Hurricane Katrina in Louisiana and Mississippi, along with other hurricanes that occurred in the southern US in 2004-05. Not surprisingly, a number of articles in this collection discuss Katrina and its aftermath, providing a variety of perspectives. Sharman Bridges Smith, Executive Director of the Mississippi Library Commission, offers the view of a chief officer of a state agency in "Observations of Ground Zero: From the Outside." She states that "walking into what remains of one of the state's largest public libraries and finding no indication a library was ever there is indescribable." She offers lessons based on what worked and what didn't in the ongoing recovery effort, such as that "prior relationships and preplanning make the difference in post-disaster recovery." Focusing on one library, Idella Washington, in "Rebuilding a High School Library Collection After Hurricane Katrina," details the efforts to reconstitute the Benjamin Franklin High School library in New Orleans after its destruction.

Recalling Coleridge's image of becalmed sailors dying of thirst in the midst of an ocean, Robert E. Skinner (Xavier University) in "'Nor Any Drop to Drink': New Orleans Libraries in the Aftermath of Hurricane Katrina" surveys the extent of the damage to several libraries in New Orleans, and provides a case study of how disaster was handled at his own library, and the many lessons learned, such as creating databases of contact information for all employees and distributing the information widely. Kay L. Wall, in "Lessons Learned from Katrina: What Really Matters in a Disaster," similarly surveys the extent of damage to several libraries in the Mississippi Gulf Coast area, and distinguishes differing responses to what she calls minor disasters, localized disasters, major disasters, and catastrophic disasters. In the latter, "assume no regular telephone service, no cell phone, little mass media, and no internet service." She counsels the creation of printed staff contact lists to respond to such a breakdown of normal communications mechanisms.

In "The 2004 and 2005 Gulf Coast Hurricanes: Evolving Roles and Lessons Learned for Public Libraries in Disaster Preparedness and Community Services," authors Paul T. Jaeger, Lesley A. Langa, Charles R. McClure, and John Carlo Bertot discuss the significant role that public libraries played in the aftermath of these events, "ensuring access to vital information" and successfully providing "a range of disaster preparedness and recovery services that were not provided and could not have been provided by other government agencies." In "SOLINET's Gulf Coast Libraries Recovery Projects for Public and Academic Libraries," Kate Nevins and Sandra Nyberg, both of whom work for SOLINET (the Southeastern Library Information Network), discuss the two major grants SOLINET is administering, on behalf of the Bill & Melinda Gates Foundation and the Andrew W. Mellon Foundation, to assist the recovery efforts of libraries in the region.

The problems discussed in this collection could happen in any library. One of them, mold, happened at the Florida Atlantic University Library (where I am Director) last summer, during the construction of a new wing to the library. Poor planning and execution involving the breaking through of the old outside wall without adequately sealing off the old wing, compounded by old and poorly-functioning air conditioning, at the height of Florida's steamy summer, exposed our book collection to the great outdoors in August, and caused a mold outbreak that cost $100,000 to address. We hope that reading the items in this collection, in conjunction with guides to planning and preparedness, will result in much more awareness on the part of all librarians, and "pre-need" thinking about what to do before disasters strike.

It is human nature to hope that "it will never happen in our library" and to convince oneself that this is certainly the case, and hide one's head in the sand, but the likeliness of an emergency caused by natural events is high, and preparation is the best defense against disaster.

William Miller
Director of Libraries
Florida Atlantic University

A Selected Bibliography of Library Disaster Stories: Before, During, and After

Jane Strudwick

SUMMARY. This selected bibliography includes international sources detailing disasters in libraries, as well as disaster planning and recovery resources. doi:10.1300/J118v25n03_02 *[Article copies available for a fee from The Haworth Document Delivery Service: 1-800-HAWORTH. E-mail address: <docdelivery@haworthpress.com> Website: <http://www.HaworthPress.com> © 2006 by The Haworth Press, Inc. All rights reserved.]*

KEYWORDS. Bibliography, disaster planning, disaster management, disaster recovery, libraries, hurricane, flood, fire

INTRODUCTION

The following selection of sources was chosen to illustrate the human capacity for creativity and resilience during trying times, and to assist the librarian engaged in disaster planning and disaster planning research. Since disasters such as 9/11, the 2004 Southeast Asia earthquake and tsunami, and the devastating 2005 hurricane season, there has been a need to once again examine the library's role in its community and its role as the protector of our cultural history.

Jane Strudwick, MLS, is Coordinator of Academic Support Services, Florida Atlantic University, S.E. Wimberly Library, 777 Glades Road, Boca Raton, FL 33431 (E-mail: jstrudw1@fau.edu).

[Haworth co-indexing entry note]: "A Selected Bibliography of Library Disaster Stories: Before, During, and After." Strudwick, Jane. Co-published simultaneously in *Public Library Quarterly* (The Haworth Information Press, an imprint of The Haworth Press, Inc.) Vol. 25, No. 3/4, 2006, pp. 7-16; and: *Dealing with Natural Disasters in Libraries* (ed: William Miller and Rita M. Pellen) The Haworth Information Press, an imprint of The Haworth Press, Inc., 2006, pp. 7-16. Single or multiple copies of this article are available for a fee from The Haworth Document Delivery Service [1-800-HAWORTH, 9:00 a.m. - 5:00 p.m. (EST). E-mail address: docdelivery@haworthpress.com].

Available online at http://plq.haworthpress.com

doi:10.1300/J118v25n03_02

The stories of libraries serving those displaced by the floods and winds of Hurricane Katrina asserted the library's role as, what Francine Fialkoff termed in her 2005 *Library Journal* editorial, "government that works," but also raised questions as to how libraries can work more efficiently during a community's recovery process and better prepare for the additional responsibilities. Many of these articles and manuals are not yet written, but we do have the instructive trial and error accounts of our colleagues as they trudged up the learning curve. In the "Stories: During the Disaster" section of this bibliography I have included some of these Gulf Coast accounts, along with other narratives detailing librarians' experiences with a variety of disasters from around the globe.

Some of the other stories included, especially from Europe, illustrate the library's role as protector of cultural history. From the pictorial retrospective of the flooding of Florence (<http://www.mega.it/allu/eng/ealluvho.htm>) to the recitation of the lost manuscripts of Bosnia-Herzegovina in András Riedlmayer's "Libraries are not for Burning" (1996), it is evident what is at stake–the echoing permanence of cultural loss. The "Recovery: After the Disaster" section of the Bibliography includes stories of restored collections and the rebuilding of damaged structures; however, as the recovery process often encompasses the persisting crisis of homelessness, loss, poverty, and governmental disconnect, these sources are also part of the disaster story itself.

The "Planning: Before the Disaster" section, like the other parts of the bibliography, includes the narratives of planning–what worked, what didn't work–as well as disaster planning guides and resources.

I hope the following resources will soon be supplemented by further stories of recovery, fortitude, resilience and ingenuity, as librarians meet the inevitable or forestall when they are able. I also anticipate that new planning resources addressing our current understanding of disaster planning and recovery, and reflecting the hard lessons recently learned, will be available in the near future. Most importantly, I hope that these fascinating and true stories will prompt librarians to reexamine their own disaster plan.

STORIES: DURING THE DISASTER

Dewan, Shaila K. "Books Spirited to Safety Before Iraq Library Fire." *New York Times,* July 27, 2003, Late Edition (East Coast).

Diamond, Tom. "The Impact of Hurricanes Katrina and Rita on Three Louisiana Academic Libraries." *Library Administration and Management* 20, no. 4 (2006): 192-200.

Dubicki, Eleonora. "Surviving the Loss of Access to Print Periodicals." *Technical Services Quarterly* 23, no. 2 (2005): 1-17.

Eberhart, George M. "Hurricane Rita Wreaks Havoc on Louisiana, Texas." *American Libraries* 36, no. 10 (November, 2005): 19-20.

Eberhart, George M., Gordon Flagg, Beverly Goldberg, Pamela Goodes, Leonard Kniffel, Greg Landgraf, and Amy Stone. "Katrina's Terrible Toll: Librarians Rally to Provide Information for a Devastated Gulf Coast Population." *American Libraries* 36, no. 9 (2005): 14-18.

Eng, Sidney. "How Technology and Planning Saved My Library at Ground Zero." *Computers in Libraries* 22, no. 4 (2002): 28-35.

Fineberg, Gail. "Surviving War: Library Sends a Team to Aid Iraqi Librarians." *Library of Congress Information Bulletin* 62, no. 12 (2003): 298-301.

Grant, Alison. "Benighted! How the University Library Survived the Auckland Power Crisis." *Australian Academic & Research Libraries* 31, no. 2 (2000): 61-68.

LeBoeuf, Mary Cosper. "Disasters Strike, Public Libraries Prevail: The Impact of Hurricanes Katrina and Rita on Louisiana Public Libraries." *Louisiana Libraries* 68, no. 4 (2006): 3-7.

——— "Ill Winds." *Public Libraries* 45, no. 3 (2006): 58-63.

McKnight, Michelynn. "Health Sciences Librarians' Reference Services during a Disaster: More than Collection Protection." *Medical Reference Services Quarterly* 25, no. 3 (2006): 1-12.

Morris, John. "The Los Angeles Central Library Fire." *Library Association Record* 88, (1986): 441-43,

Pintus, Sandro and Silvia Messeri. "Florence 30 Years after the Flood." *Florence Art News* 1, no. 1 (1996): January 5, 2007, <http://www.mega.it/allu/eng/ealluvho.htm> (accessed January 20, 2007).

Price, Peggy. "Depends on What You Mean by 'Disaster.'" *Mississippi Libraries* 70, no. 3 (2006): 56-57.

Puckett, Rahye and John L. Whitlock. "Katrina Blew in and the House did Not Fall." *Mississippi Libraries* 70, no. 3 (2006): 60-61.

Ray, Emily. "The Prague Library Floods of 2002: Crisis and Experimentation." *Libraries and Cultural Record* 41, no. 3 (2006): 381-91.

Saunders, Margaret. "How a Library Picked Up the Pieces After IRA Blast." *Library Asscociation Record* 95, no. 2 (1993): 100-101.

Sturges, Paul and Diana Rosenberg, eds. *Disaster and After: The Practicalities of Information Service in Times of War and Other Catastrophes.* Proceedings of an International Conference Sponsored by IGLA (the International Group of the Library Association), 4-6 September 1998, University of Bristol. London: Taylor Graham, 1999.

Taylor, Dee Dee. "Hurricane Survival Tactics." *American Libraries* 35, no. 10 (2004): 42-43.

Wilkins, Katherine. "Hindsight: The Last Day of Smith." *American Libraries* 37, no. 7 (2006): 40.

Yamamoto, Bazzell Tokiko. "Halloween Eve Flash Flood Disaster in Paradise: Disaster Recovery Report for Hamiliton Library of the University of Hawaii at Manoa." *Joho Kanri* 48, no. 6 (2005): 356-365.

Young Jr., Terrence E. "If I Only Knew then . . . ." *Library Media Connection* 24, no. 7 (2006): 22-23.

STORIES OF RECOVERY: AFTER THE DISASTER

Alire, Camila A. "The Silver Lining: Recovering from the Shambles of a Disaster." *Journal of Library Administration* 38, no. 1/2 (2003): 101-107.

Amarasiri, Upali. "Rising from the Wreckage: Development of Tsunami-Affected Libraries in Sri Lanka." *IFLA Journal* 31, no. 4 (2005): 307-14.

Amigos Library Services. "Imaging and Preservation Service." http://www.amigos.org/preserve.html#disaster> (accessed January 2, 2007).

Atkins, Winston and Ellen Belcher. "Coordinating a Bomb Blast Recovery." *Conservation Administration News* 55, (October 1993): 1-2, 24-27.

Balik, Vojtech and Jiri Polisensky. "The National Library of the Czech Republik and the Floods of 2002." *Alexandria* 16, no. 1 (2004): 17-24.

Cervone, H. Frank. "Disaster Recovery and Continuity Planning for Digital Library Systems." *OCLC Systems & Services* 22, no. 3 (2006): 173-78.

Chadbourne, Robert D. "A Post-Disaster Primer: Elba on the Rebound." *Wilson Library Bulletin* 68, (May, 1994): 24-25.

Clareson, Tom and Jane S. Long. "Libraries in the Eye of the Storm: Lessons Learned from Hurricane Katrina." *American Libraries* 37, no. 7 (2006): 38-41.

Curzon, Susan Carol. "When Disaster Strikes: The Fall and Rise of a Library." *American Libraries* 31, no. 4 (2000): 64-69.

Cveljo, Katherine. "Wounded Libraries in Croatia–Destruction and Heroic Recovery Efforts." *International Leads* 12, no. 4 (1998): 1-4.

Dearstyne, Bruce W. "Taking Charge: Disaster Fallout Reinforces RIM's Importance." *Information Management Journal* 40, no. 4 (2006): 37-43.

Dickerson, Lon R. "Building Even Better Libraries, Post-Katrina." *American Libraries* 36, no. 10 (November, 2005): 16-18.

Eberhart, George M. "New Orleans Deals with Layoffs, Damage." *American Libraries* 36, no. 10 (November, 2005): 18-19.

Furlong, Scott R. and Denise Scheberle. "Earthquake Recovery: Gaps between Norms of Disaster Agencies and Expectations of Small Businesses." *The American Review of Public Administration* 28, no. 4 (December, 1998): 367-389.

Hill, Roger. *The Restoration of Books: Florence 1968.* Royal College of Art, London, 1968; 39 min.; 1 videodisc. Mounted by the University of Utah as a free streaming video at <http://data.scl.utah.edu/fmi/xsl/stream/details.xsl?-recid=354> (accessed February 5, 2007).

Jefferson, Theresa L. "Knowledge/Information Technology Engineering Management: Evaluating the Role of Information Technology in Crisis and Emergency Management." *VINE: The Journal of Information & Knowledge Management Systems* 36, no. 3 (2006): 261-264.

Jones, Calvert and Sarai Mitnick. "Open Source Disaster Recovery: Case Studies of Networked Collaboration." *First Monday* no. 5 (2006). Accessible from <http://www.firstmonday.org/issues/issue11_5/jones/index.html>.

McEntire, David A. *Disaster Response and Recovery*. Hoboken, NJ: John Wiley & Sons, 2007. See esp. chap. 2, "Understanding the Actors: Role and Responsibility of Disaster Participants," and chap. 7, "Managing Media Relations, Donations and Volunteers: Expected Challenges and Benefits."

McNutt, John G. "Electronic Government, the Internet, and Disasters: An Emerging Relationship." *DttP: A Quarterly Journal of Government Information Practice & Perspective* 34, no. 1 (2006): 17-21.

Paton, Douglas and Rhona Flin. "Disaster Stress: An Emergency Management Perspective." *Disaster Prevention and Management* 8, no. 4 (1999): 261-267.

Rachmananta, Dady P. "Rehabilitation and Reconstruction Plan for Libraries and Documentation Centers in the Province of ACEH." Open Seminar on the Documentary Heritage Damaged by the Indian Ocean Tsunami, National Diet Library, Tokyo, December 6, 2005. <http://www.ndl.go.jp/en/iflapac/pdf/rachmananta.pdf> (accessed January 20, 2007).

Riedlmayer, András. "Libraries are not for Burning: International Librarianship and the Recovery of the Destroyed Heritage of Bosnia and Herzegovina." *Inspel* 30, no. 1(1996): 82-94.

Sakamoto, Isamu. "One Step Forward–Relief Work for Damaged Documents at Aceh, Indonesia." Open Seminar on the Documentary Heritage Damaged by the Indian Ocean Tsunami National Diet Library, Tokyo, December 6, 2005. <http://www.ndl.go.jp/en/iflapac/pdf/sakamoto.pdf> (accessed January 20, 2007).

Su Pinhong and Lin Shitian. "The Urgent Need to Preserve and Conserve Ancient Books: From the Event that Ancient Books Collected in Yongquan Monastery were Damaged by a Flood: Case Study." *International Preservation News* no. 38 (2006): 15-21.

Swartz, Nikki. "Dealing with Disaster." *Information Management Journal* 40, no. 4 (2006): 28-34.

PLANNING: BEFORE THE DISASTER

Adinku, Sarah. "Towards Disaster Preparedness and Recovery Planning Procedures for Libraries: A Survey of Staff and Users of the Balme Library, University of Ghana." *African Journal of Library, Archives & Information Science* 15, no. 1 (April, 2005): 75-79.

Alire, Camila, ed. *Library Disaster Planning and Recovery Handbook*. New York: Neal-Schuman, 2000.

Boice, Daniel. "Disaster Planning." *Catholic Library World* 76, no. 3 (2006): 245-245.

Champion, Sandra and Christine Master. "When Disaster Strikes." *School Library Journal* 39, no. 9 (1993): 146-149.

Conversation OnLine. "Disaster Preparedness and Response." <http://palimpsest.stanford.edu/bytopic/disasters/> (accessed 19 December 2006).

DePew, John, N. "Statewide Disaster Preparedness and Recovery Program for Florida Libraries." *University of Illinois Graduate School of Library and Information Science: Occasional Papers* 185, no. February (1989).

"Disaster Help." *Medicine on the Net* 12, no. 3 (2006): 18.

Edwards, Bruce and John Cooper. "Testing the Disaster Recovery Plan." *Information Management & Computer Security* 3, no. 1 (1995): 21-27.

Georges, Jo-Ann. "Skills Development and Management for Disaster Mitigation Planning: The Specific Case of Electronic Equipment and Digital Data." *International Preservation News* no. 34 (2004): 12-14.

Graham, Kristy and Dirk H. R. Spennemann. "State Emergency Service Local Controllers' Attitudes Towards Disaster Planning for Cultural Heritage Resources." *Disaster Prevention and Management* 15, no. 5 (2006): 742-762.

Halsted, Deborah, D., Richard Jasper P., and Felicia Little M. *Disaster Planning*. New York: Neal-Schuman, 2005.

Hiatt, Charlotte J. *A Primer for Disaster Recovery Planning in an IT Environment*. Hershey, PA.: Idea Group, 2000.

Kahn, Miriam. *Fires, Earthquakes and Floods: How to Prepare Your Library and Staff. ONLINE* 18, (May 1994): 18-24.

———. "Mastering Disaster." *Library Journal* 118, no. 21 (1993): 73-75.

———. *Protecting Your Library's Digital Sources*. Chicago: ALA, 2004.

———. *Disaster Response and Planning for Libraries*. Second ed. Chicago: ALA, 2003.

Matthews, Graham and Paul Eden. *Disaster Management in British Libraries: Project Report with Guidelines for Library Managers*. Library and Information Research Report 109. London: The British Library Board, 1996.

Matthews, Graham and John Feather. *Disaster Management for Libraries and Archives*. Hampshire, England: Ashgate, 2003.

Matthews, Graham. "Disaster Management: Sharing Experience, Working Together Across the Sector." *Journal of Librarianship & Information Science* 37, no. 2 (2005): 63-74.

Muir, Adrienne and Sarah Shenton. "If the Worst Happens: The Use and Effectiveness of Disaster Plans in Libraries and Archives." *Library Management* 23, no. 3 (2002): 115-123.

National Fire Protection Association. *NFPA 909: Code for the Protection of Cultural Resources Properties–Museums, Libraries, and Places of Worship, 2005 Edition*. Also available online at <http://www.nfpa.org/freecodes/free_access_document. asp?id=90905>.

Oliver, Marshall. "A Librarian Copes with Disaster in New Orleans." *Library Media Connection* 24, no. 7 (2006): 18-21.

Robertson, Guy. "Water Finds a Way: Dealing with Leaks and Floods in Your Library." *Feliciter* 51, no. 2 (2005): 83-85.

Scott, Jude and Johanna Wellheiser. With assistance from John Barton. *An Ounce of Prevention: Integrated Disaster Planning for Archives, Libraries, and Record Centres*. Lanham, MD: Scarecrow Press, 2002.

Shorley, Deborah. "Disaster Planning 'in the End You Just Cope.'" *Library & Information Update* 2, no. 3 (2003): 46.

Thenell, Jan. *The Library's Crisis Communications Planner: A PR Guide for Handling Every Emergency*. Chicago: ALA, 2004.

Trotta, Carmine J. and Marcia Trotta. "Preparing Emergency and Disaster Plans." Chap. 7 in *The Librarian's Facility Management Handbook*. New York: Neal-Schuman Publishers, 2001.

SOURCES FOR SAMPLE DISASTER PLANS

Conversation OnLine. "Disaster Plans." <http://palimpsest.stanford.edu/bytopic/disasters/plans/> (accessed December 19, 2006).

"dPlan: The Online Disaster-Planning Tool." Prepared by the Northeast Document Conservation Center (NEDCC) and the Massachusetts Board of Library Commissioners (MBLC).<http://www.dplan.org/> (accessed December 28, 2006).

Emergency Planning and Management in College Libraries, compiled by Susan C. George. CLIP note, no. 17. Chicago: Association of College & Research Libraries, 1994.

"The Robinson Library Disaster Control Plan." <http://www.ncl.ac.uk/bindery/disaster.html> (accessed 02 January 2006).

WEBSITES

American Red Cross Disaster Services Guide

<http://www.redcross.org> (accessed 1/25/07)

The American Red Cross website's disaster pages offer valuable information on preparation, disaster safety and recovery, and an office locator by zip code on the homepage. For patrons and librarians, the breakdown of services by specific disaster and the Community Services page are important features. The *Safe and Well* portion of the website provides victims of disasters a way to register, allowing their friends and family to know that they are "safe and well." The site is user friendly and all of the messages are standardized. Search results provide the loved one's first and last name, date of post, and the standard message selected.

Federal Management Agency Homepage (FEMA)

<http://www.fema.gov/> (accessed December 18, 2006)

The FEMA site is a one-stop resource for planning and recovery, and a source for federal forms, applications and brochures. The site is conveniently organized to target the specific needs of its users: Business, Emergency Personnel, Governments, Individuals, Institutions, Kids, News Media, and Professionals. An important aspect for librarians is the "Information Resource Library," a searchable database of all publicly accessible FEMA information and resources.

The International Committee of the BlueShield (ICBS)

<www.ifla.org/blueshield.htm> (accessed December 7, 2006)

The ICBS is the cultural equivalent of the Red Cross. By bringing together the expertise of international cultural organizations, the ICBS advises and assists museums and archives, historic sites, and libraries to recover from armed conflict or natural disaster events. The site offers a global perspective on cultural sites at risk by providing links to bulletins, reports, publications, developments, conferences, and statements and appeals.

Inland Empire Libraries Disaster Response Network (IELDRN)

<http://www.ieldrn.org/MissHist.htm> (accessed December 27, 2006)

The IELDRN is composed of academic, community, and public libraries from the Inland Empire Region of California and was formed to pool disaster recovery and planning resources. This website is a good example of a library disaster response network and includes a sample disaster plan and model mutual aid agreements. Also see "San Diego/Imperial County Libraries Disaster Response Network (SLIDRN)" <http://orpheus.ucsd.edu/sildrn/> (accessed December 19, 2004) for more information on regional networks.

Disaster! Finder: A Search Service Provided by the NASA/GSFC Earth-Sun Exploration Division

<http://disasterfinder.gsfc.nasa.gov/Disaster_Management/> (accessed December 19, 2006)

A comprehensive tool designed to quickly locate the latest links in disaster information. Over 600 sites from the disaster community are indexed. All of the major organizational sites encompassing disaster management and mitigation, recovery, relief, research, and response, as well as the major disciplines and workshops and conferences are arranged in an easily negotiated taxonomy.

Disaster Mitigation & Recovery Resources

<http://www.solinet.net/preservation/preservation_templ.cfm?doc_id=71> (accessed January 2, 2007)

The Southeastern Library Network (SOLINET) offers publications, leaflets, videos, and news on preservation and hurricane preparedness.

Many of the resources are also offered in Spanish and are available online. The SOLINET Preservation Field Service Program also offers telephone emergency assistance for institutions and individuals.

Northeast Document Conservation Center (NEDCC)

<http://www.nedcc.org/home.php> (accessed December 28, 2006)

The NEDCC is a non-profit regional center for the conservation and preservation of paper-based materials. Made possible by a grant from the National Endowment for the Humanities (NEH), the NEDCC provides a 24/7 emergency phone assistance service for cultural institutions whose collections are threatened by disaster. It is also useful for patrons with inquiries about the conservation and preservation of their private collections.

BIBLIOGRAPHIES

Amigos Library Services. "Emergency Preparedness and Recovery: A Selected Bibliography." <http://www.amigos.org/preservation/bibdis.html> (accessed December 27, 2006).

"The Bosnian Manuscripts Ingathering Project." See the home page, <http://www.kakarigi.net/manu/ingather.htm> and the "More Information" page, <http://www.kakarigi.net/manu/moreinf.htm> (accessed January 15, 2007).

Since the May 1992 burning of the Oriental Institute in Sarajevo a group of American and Canadian scholars, in concert with librarians from the Oriental Institute, began "The Bosnian Manuscript Ingathering Project" initiative to recover and reassemble the lost and destroyed materials. This site includes lists, websites, and articles chronicling the losses and ongoing recovery efforts to recover the cultural heritage of Bosnia-Herzegovina.

Kulzack, D. and L. Lennertz. "A Decade of Disaster: A Selected Bibliography of Disaster Literature, 1985-1995." *Library and Archival Security* 15, no. 1 (1999): 7-66.

Murray, Toby. *Bibliography on Disasters, Disaster Preparedness and Disaster Recovery.* Tulsa, OK: University of Tulsa, 1990.

Smith, Lynn. "Disaster Recovery Resources." *MAC Newsletter* 34, no. 2 (2006): 29-30.

doi:10.1300/J118v25n03_02

Coming Back From Major Disaster: Month One

Susan Carol Curzon

SUMMARY. This article explores the most important tasks, approaches, and strategies necessary during the first month after any major disaster. Working with staff in emotional stress, working with the incident command center, surveying the damaged library, communicating with staff, key steps in restoration, and important management considerations are all discussed. doi:10.1300/J118v25n03_03 *[Article copies available for a fee from The Haworth Document Delivery Service: 1-800-HAWORTH. E-mail address: <docdelivery@haworthpress.com> Website: <http://www.HaworthPress.com> © 2006 by The Haworth Press, Inc. All rights reserved.]*

KEYWORDS. Emotional trauma, Incident Command Center, surveying damage, communication, restoration, management of disaster, attitude

---

Susan Carol Curzon, PhD, Dean, Oviatt Library, California State University, Northridge, 18111 Nordhoff Street, Northridge, CA 91330-8326 (E-mail: susan.curzon@csun.edu).

Since 1992, Susan Carol Curzon has been Dean of the Oviatt Library for California State University, Northridge–a comprehensive, urban university with a student enrollment of 33,000.

Sue was also the Director of Libraries for the Glendale Public Library in California and also served for sixteen years in various positions for the County of Los Angeles Public Library. She also has corporate library experience.

[Haworth co-indexing entry note]: "Coming Back From Major Disaster: Month One." Curzon, Susan Carol. Co-published simultaneously in *Public Library Quarterly* (The Haworth Information Press, an imprint of The Haworth Press, Inc.) Vol. 25, No. 3/4, 2006, pp. 17-29; and: *Dealing with Natural Disasters in Libraries* (ed: William Miller and Rita M. Pellen) The Haworth Information Press, an imprint of The Haworth Press, Inc., 2006, pp. 17-29. Single or multiple copies of this article are available for a fee from The Haworth Document Delivery Service [1-800-HAWORTH, 9:00 a.m. - 5:00 p.m. (EST). E-mail address: docdelivery@haworthpress.com].

Available online at http://plq.haworthpress.com

doi:10.1300/J118v25n03_03

INTRODUCTION

Any library at any time may experience a natural disaster such as fire, flood, earthquake, hurricane, or tornado. Although the devastation to the library can take many forms, from ruined buildings to burnt materials or from waterlogged books to missing roofs, these tragic experiences share one very special feature in common. The first month after any disaster is the most critical period of time. It is during that time that you will lay the groundwork for the recovery of the library and library services, set the course for rescue and restoration efforts, and make key decisions that will impact the future of the library for many years ahead.

But how is it possible to determine a course for the library in so short a period of time? What do you do to guarantee that your library is set back on the right course? How can you cope with a disaster of this magnitude? How can you start to build back what has taken sometimes decades to create? At first, all of the responsibilities ahead will seem overwhelming in the face of a disaster that may have destroyed your building and your collections. But all of the responsibilities ahead don't have to be done in the first month. Instead, it is important to focus only on the key elements that will be the foundation for a good future for the library.

This article explores those key elements–the responsibilities, duties, and approaches–that will help you to set your library back on its path to restoration.

It is important to note though that these elements are not linear because disasters are not linear. Chaos reigns during the first month of a disaster of any magnitude so you must adjust your tasks and approaches as circumstances dictate.

This article is based upon my direct personal experience as the head of a library that suffered a major natural disaster. In 1994, the Oviatt Library at California State University, Northridge was devastated by an earthquake with a 6.7 magnitude.The Library sustained major structural damage resulting in the eventual demolition of fifty percent of the building. The collection was thrown to the floor, water-damaged, and attacked by mold. Eventually, most of the items in the collection had to be cleaned of asbestos. During the six years that it took to fully restore the Library, part of the staff, the services, and the collections existed in trailers and other temporary structures. Having also spent nineteen years in public library management prior to being in an academic library, I have drawn upon my earthquake experience and my knowledge of public libraries to develop this article.

TALKING ABOUT PEOPLE

Let's talk first about people. At the heart of every disaster is the impact on people and their lives. Anyone in a disaster situation must be prepared for the fact that everyone will have their lives seriously disrupted. On the extreme, you may lose colleagues or they may lose family members. Colleagues might have sustained serious injuries. They may be searching for loved ones and for loved pets. They might have lost homes. Many might be without water or food. There may be a breakdown in civil society. At the very least, many people will have to cope with damage to homes, loss of utilities, no electronic communication, and ruined or uncertain transportation systems.

Anyone supervising a disaster in a library has to recognize and cope with the inevitable fact that many staff from the library may simply not be available. People must secure their home life first of all. However pressing the library's needs may be, the home situation is more pressing and must be attended to first. When we talk about people, the main message in the management of a disaster is that people, and their families, are the priority. Buildings and objects, however precious, must take second seat.

This brings us to an important point. The library director, or other top managers, also might not be in a situation to come to work. The normal process of the management of the library may be suspended. People who are normally in charge might not be available for any number of reasons. While the best developed plans about disasters usually refer to the normal command structure of the library, it may be that the normal sequence of command is not possible so anyone who is the next highest ranking must take up the reins. This is why it is so important to train across the entire organization and why it is so important that everyone knows the plan and has the ability to contact colleagues.

Remember too that immediately after a disaster, and for a long time thereafter, most people will suffer emotional trauma. Fear of death and bodily harm and the grief of loss of all that was comfortable, known, and loved are powerful emotions that will govern behavior. Feelings will be taut and tempers may flare as people struggle to cope with shock. The stress from a major disaster takes a formidable toll on the body, mind, and spirit. During this time, everyone will have to reach into themselves for that extra ounce of patience, kindness, and compassion for the occasionally erratic and unpredictable behavior of colleagues. How best to handle it? It is best to talk about it early and often so everyone knows they are not alone and that their colleagues will understand. If professional counselors do eventually arrive, you should always avail yourself of them for yourself and the staff but generally there are so few of them and so many people caught up in the disaster that the chance of significant professional help is slim.

One footnote on emotional stress. You cannot predict who will be victimized the most by the stress of the disaster. Someone who normally presents a cool demeanor might be panicked and someone who is ordinarily flighty might be absolute bedrock throughout the disaster. It all depends upon their deepest self, their experiences, and the impact of the disaster on their lives. It also depends upon memories that the disaster might have awakened about previous losses and traumas. People's behavior is governed by context as well as their personal characteristics and the context of a disaster is not like anything most people have ever experienced. Anyone managing a library in disaster must simply expect that people might not be what they usually are and must give them the space to find their footing again. Having said that, you will be moved by the ability of people to rise to the occasion and to sacrifice for the good of the library.

When we talk about people in disasters, one characteristic will rule–unpredictability. There will be the unpredictability of fate, of availability, of emotions, of behavior. It is best to recognize and remember this reality throughout any disaster.

WORKING WITH THE INCIDENT COMMAND CENTER

One of the first steps for whichever person or group of persons who are now in charge of the library is to report to the Incident Command Center. Almost every municipality has an Incident Command Center which is often located near the police or fire station. The Incident Command Center is the place designated and created by the municipality as the center for disaster management. Coming to the Incident Command Center will be the administration of the municipality, its elected officials, department heads, police, fire, and any other emergency personnel. The press will also converge on this center. This is the first place for you to report to ascertain the scale of the disaster. In fact, the municipality might require staff to report to the Incident Command Center first.

Why is this so often a requirement? First, the municipality, in the form of the Incident Commander, also wants to determine the availability and the condition of personnel. Second, the Incident Commander may have very specific directions for you. Third, the Incident Commander wants to control access to buildings and to movement around the city. At this time, you are naturally consumed by your interest in the library and its fate but from the perspective of the municipality, you are a part of a whole, part of the total personnel available to cope with the disaster. In other words, you are not in a vacuum. You are expected to be a team player and to put the greater needs of the municipality first.

Do remember these priorities that will govern the municipality's approach. Its priorities are: life-saving activities, saving properties, restoring law and order, restoring services such as water and heat, securing help from communities which can send assistance, providing shelter and food, getting the message out about the scale of the disaster, working with press, and initiating contact with the federal government for FEMA funding. If the library fits into those priorities, it will receive attention. If not, the library will not be the focus for some time to come.

Remember also that in a disaster, the management style may change to a quasi-military one because of the emergency situation. People trained to manage disasters such as police and fire chiefs may be giving you orders and expecting them to be carried out. Your goal? Be helpful and be part of the team. Also, remember that the Incident Commander or chiefs may themselves be traumatized by the disaster and therefore must also be given space as they try to restore the municipality.

Incident Command Centers are also the center of politics. During this time, the elected officials, and upper administrators, want to demonstrate that they are leaders. Their eye is on both the restoration of the municipality and also on their own future, especially once the press is involved. Many people both rise and fall not long after a disaster depending upon their behavior and the circumstances. Eventually, there is always blame about how the disaster occurred or the municipality's preparation or the management of the disaster. It is rare that someone or a group of someones do not take the fall. Librarians with wisdom fly below this radar screen and focus on the task of restoring the library. In short, stay out of the politics and do your job.

Be prepared also that your library, if it is in relatively good condition, may be requisitioned as a shelter, hospital, or tragically a morgue. Because of prior emergency planning, you may know how your library might be used but there are many decisions in the moment that can cause the library to be used for many things. Again, it will be an issue of priorities and no one will prioritize library resources over human needs. If the library is requisitioned, most likely the library's staff job will be to make space available accordingly and with as little damage as possible to the resources of the library. Again, if you know ahead of time how your library might be used in a disaster, it is worth having a plan that details how the new function will be accommodated rapidly.

Now if you are not needed by the Incident Commander, you should still keep returning to the Incident Command Center once or twice a day for vital information. Don't be left out of this loop. There will be regular meetings and briefings of which you will need to be a part. Importantly, this is the most likely place where you will be able to bring up the needs of the library and to seek approval for actions that you may need to take.

SURVEYING THE DAMAGED LIBRARY

Of course, your other immediate priority will be to survey the damage to the library even if you were in the library when the disaster struck and had to evacuate the building. In this task, you must be very careful. Many dangers lurk in damaged buildings. Buildings can look structurally sound but may not be. Inside, there may be hazardous materials conditions, asbestos, loose live wires, and objects or glass and tile still waiting to fall.

Of course, don't go alone and make sure that someone at the Incident Command Center knows that you are going to the library. Take your direction from the Incident Commander about whether or not any entry to buildings is permitted. A damaged building is very unpredictable so you will understand if the Incident Commander is very cautious. Also, when the municipality is in crisis, emergency personnel don't need to be rescuing yet another person.

However, if you are able to move forward to survey the damage to the library, follow these steps even if the damage seems mild. First, traverse at a safe distance the outside of the building. Look up at the roof and at any pillars or windows. Look down to see what is happening on the ground. Then, look carefully in the windows. Watch for cracked ceilings, broken windows, and cracked floors. Use all your senses. Hear what the building sounds like. Smell for unusual odors. Remember to use your intuition as well as your reasoning to determine a building's safety. We aren't engineers so we can be easily fooled. Even engineers do not know the safety of a building at a glance. If the library isn't safe or you don't feel it is safe, don't go in. No one will expect that of you. Wait for the emergency personnel or engineers to go with you as determined by the municipality. Foolish courage helps no one.

Before you start on your journey to survey the library, be sure to dress the part even if you are only looking at the outside. You might feel silly but you will be glad you have the right gear. Equip yourself with heavy shoes, protective clothing, and a hard hat. Carry a small backpack equipped with water, a flashlight, gloves, whistle, your identification and emergency numbers, safety glasses, a facemask, and some dried food. Be sure to carry some small tools such as a penknife, wrench, hammer, and screwdriver in case you need to open doors or drawers. Take a notebook because you may need to remember some points or draw some diagrams of damage for the engineers. Don't forget your camera because documenting must start right away.

Remember also that there rarely is only one disaster. Tragically, most disasters have a sequel. Floods can return, the earth can keep shaking, and fires can backtrack or burn unseen causing later explosions. Be mindful of this two-part reality of disasters when you are out and about in the first day or so of a disaster. In other words, continue to be wary. The world after a disaster of any proportion is not a world most of us know and therefore you must be as watchful as if you had landed on an alien planet.

Now, in libraries that have sustained major disasters, there is always one guaranteed part two. Molds, those opportunistic fungi, are particularly fond of libraries that have fallen victim to disaster. Every library disaster plan should also include a plan for the detection, prevention, and early eradication of molds.

When you are done surveying, write up everything that you noticed that is of concern to you. This information will help you to plan the restoration effort and will provide essential information to the municipality's security personnel who must protect the library's assets and to the engineers as they determine a building's fate.

COMMUNICATING WITH STAFF

Another vital task in any disaster is to communicate with your colleagues. First, you want to find out how their situation is and find out who needs help. Second, you want to know who is available to help other colleagues or to help at the library or the municipality.

The challenge is that normal modes of communication are impacted greatly in a disaster. Very likely phone and email will be down and it may be days before they are restored again. That is why is it useful to have some designated meeting places so that, if people can get there, some communication can occur.

However, as soon as possible, the phone tree should be activated to pass on and to receive information. For this reason, the managers and supervisors of the library should have, in their office and in their home, the phone numbers, home and cell, email, and addresses of all personnel. The staff should have the phone numbers of the managers and supervisors so they can also make contact. As part of disaster planning, the phone tree, that follows the line of the organization, should have been established so that one person is not making many calls but that many are engaged in the task. You will find that many staff who have not yet had the "return to work" call will be willing to be engaged in passing on messages about the library's situation or about each other's welfare.

Assuming that your staff is okay, which is a big assumption in a major disaster, one of their first preoccupations will be whether or not they still have a job. This of course is not a decision that you make alone. The return to work of all available personnel is usually the decision of the municipality. Of course, all the emergency personnel return to work, without asking, along with other department heads and their chief deputies. Library personnel, other than the managers, would not ordinarily return immediately in a disaster and, in general, it is best that most people stay home and out of the way of emergency work.

If the library is not deemed a priority for restoration, there is a possibility that library employees might be laid off. If you have any impact on this discussion, you can use these two arguments. First, rescuing library resources, which often represent a cumulative investment of millions of dollars for any municipality, is vital. Library employees in turn are vital for that rescue effort. Second, library employees are versatile and can be deployed into many other positions temporarily for the emergency services. In particular, library staff can take up the responsibility of a municipal emergency information desk. This is a service which every municipality needs as the calls flood in from everywhere in the community seeking information about everything from lost pets to downed trees. An emergency information desk is a function at which library staff especially excels.

If your staff is temporarily laid off or working elsewhere in the municipality, still it is important to continue to communicate. Staff doesn't stop being library staff just because they are elsewhere and you still want to have an intact organization when they eventually return to work at the library. Continue to activate the phone tree with important information and send out emails as soon as that is restored to keep staff up to date on the library's progress. Don't hesitate also to share the personal news as appropriate as people will care very much about the status of their colleagues. If the library takes a very long time to restore, be prepared that you may lose some staff to other departments over time as staff get interested in new work and as those departments become interested in talented library staff.

FIRST STEPS IN RESTORATION

There are three vital tasks that must occur after a disaster. First, the building must be secured. Vandalism is one of the tragic aftermaths of disasters so a library with broken windows and broken doors will be tempting to those who are predatory during a disaster. If possible, and sometimes the damage is so bad that it is not possible, the library should be secured. Here, ordinarily, you would work with the municipality's facilities personnel to board up or otherwise secure the library. That security can take the form of fencing or boarding up or even actual security personnel stationed around the library so access to the library is controlled. You must be a strong voice with the municipality's leaders about the importance of securing the library. However, be prepared that you may have to face the reality that the library will not be an immediate priority.

The second vital task is the rescue of the collection. There are so many decisions to be made about the collection depending upon the nature of the disaster and its impact on the collection. Were the resources damaged by fire? Flooding? Thrown to the floor? Scattered inside and around the building? Have the shelves collapsed, making rescue difficult? Is the building open to the elements allowing water, wind, and insects in? What parts of the collection should be rescued first? Again, rescuing the collection shows the importance of having a disaster plan that contains collection rescue priorities, the plan of action for the rescue, and emergency contact numbers of material restoration companies.

Once again you must be a strong voice with the municipality's administrators to allow you, the library staff, and restoration companies in so that you can move rapidly to rescue the collection if it is safe to do so. For library staff, rescuing the collection will be the most important, absorbing, and time consuming task. Make sure though that you are organized with the rescue effort because much time and labor can be lost without clear directions and course of action. Of course, when embarking on the rescue effort, don't forget that staff should be equipped for the task with masks, gloves, safety glasses, and protective clothing. Watch out for exhaustion of the staff because they will work very hard until the collection is completely rescued.

Also, don't hesitate to ask for volunteers from the community because people have very warm feelings about libraries and will be very willing to help out. However, remember that volunteers need supervision so be careful not to get too many people too fast and beyond the ability of the library to handle too many volunteers.

Now, after the building is secured and collection rescue operations underway, the next step is to plan how you will restore library services. If your library is badly damaged and out of commission for some time, remember that you will have to go down two tracks simultaneously in the restoration of services. On the one hand, you will commence planning how to rebuild the library. On the other hand, you will commence planning how to offer services in alternative space until the library is restored. Will you be able to operate services from a storefront? From trailers? Much of this decision relating to be space will be governed by the municipality's plans and that brings us to an important point. After a disaster, good space is hard to come by. Make sure you are at the table when the discussions of allocation of space occur. Emergency services go first of course as will municipal administration but the library had needs too and, once again, you must be a strong voice for the library.

Of course, there may be some services that the library can provide without waiting on the municipality. For example, how fast can you restore your electronic resources and services? Are there additional electronic services that can be provided? What about extending the services of your bookmobile if it hasn't been requisitioned for other use? What about books-by-mail services? What services can your neighboring libraries provide to your community especially if you send your personnel to work there to help the people of your community?

Obviously during the first month, major restoration of the building will not occur. However, preliminary discussions may occur sooner than you might think. Again, you will want to be at the table making sure that the library is in the schedule for restoration. You will need to establish yourself, by both presence and preparation, as pivotal in the rebuilding of the library. One would think it would be obvious that the library's manager would be pivotal in the rebuilding but decisions in the aftermath of a disaster do not always follow predictable paths. In particular, you have to watch out for opportunists. For example, if your library sits on a prime location and is a nice building, some who have had their eye on the place or the location for a long time might take this opportunity to challenge whether the library should be restored there or even if a library is needed at all in the digital age. You may not experience these extremes but don't be surprised if off-the-wall discussions occur. Many normal controls are off after a disaster and there can be a space grab. Use cool logic in the face of these discussions, regardless of how frustrating the conversation is, but be firm about the library's mission and importance.

On the other hand, don't be closed to a very good new idea. It is possible that the municipality has long had plans that would be better for the library and the community. There is no better time to make changes than after a disaster when nothing is as it was.

The third most important task is documentation. You must start documenting immediately in preparation for the FEMA funding which will eventually come. Keep a daily diary of actions. Keep logs of all expenses, large and small, including personnel costs. Document the library's changing situation through still photos and videos. Keep duplicates because once these records are turned over to FEMA, you will not see them again. Assign this daily recording to staff that may not be able to be involved in other rescue operations.

In time, the municipality will give you directions as to what records to keep and how to keep them. However, often the first month can go by without the bureaucracy kicking in. You don't want to lose those first month expenses just because the municipality has not yet told you how to proceed.

Remember that the FEMA team does not come one time only. They come throughout the disaster and sometimes challenge what their own teams have decided before. You may have to go before the FEMA team and defend your decisions, your expenses, and even the purpose and function of a library. Therefore, make sure that you have the best record possible of all actions, decisions, and expenses.

SOME MANAGEMENT CONSIDERATIONS

Several times throughout this article, the importance of you as a strong voice for the library has been brought up. There is an enormous amount of "noise" after a disaster. Everyone has needs. Everyone is trying to get their agenda across. Everyone has ideas and opinions. You must also be heard. This is not a time to sit back. Obviously, you know that the library cannot be first before emergency or life-preserving services or the restoration of utilities. However, the library cannot be at the end of the line either but that is where it will be unless a strong voice for its needs is heard.

Another management consideration is to avoid one very powerful impulse after disaster and that is to run around and try to do everything immediately. Instead, think, think, think before you act. Consider the consequences of the decisions that you make. Some of the decisions that you make may affect actions for years in the future. Think downstream. How will this decision play out? And, of course, don't think alone. Talk to your staff and get their input.

Also, don't micromanage in a disaster. Of course, no one will expect extensive committee meetings and consultation. Action is the watchword in disaster so management will be more command and control than most library staff are accustomed to. However, once the direction is given, let the staff run with it and don't look over their shoulder continually. First of all, you have to have your mind uncluttered by detail so you can see the big picture. Second, the amount of detail in rescuing a library is so great that you will soon be exhausted by it unless others are doing their share.

Be sure not to neglect press relations. For the first few days, the press is very interested in the disaster. A library in disaster is very compelling, and local and national press may want to speak with you. Check with the municipality's press relations first because they may have certain messages they want to get out or they may want to control information initially. If you are speaking with the press, be sure that you always have some interesting figures at your fingertips such as how the library is used by a thousand people a day or that 10,000 kids have library cards. You will also want to have sound bites so that you can convey to the press, in the most powerful way, what losing a library will mean to the community and especially to children.

Be sure also to let your colleagues in the world of libraries know what is happening. Let the major journals and listservs know what is going on. Help and advice may come as well as much needed sympathy.

Lastly, it is important to stay in touch with your boss. The normal sequence of business activity is fractured along with the buildings so any regular meetings that you might have with your boss are unlikely to be sustained. However, your boss is still pivotal even if terribly pre-occupied with larger tasks. Stay in contact, consult, seek direction, and seek approval. Eventually, the organization comes back as it once was and you want to make sure that you have preserved this vital relationship.

GETTING THE RIGHT ATTITUDE

A disaster of any scale requires you to be a long distance runner. Bringing a library back from major disaster is not a short sprint. This major task, probably the most significant in your library life, will require energy, fortitude, patience, and skill. Just remember these things.

Be sure to take care of yourself. First, keep yourself physically fit. In a disaster, the temptation is great to work night and day without mercy. However, if you really want to restore the library, you must be fit to go the distance. Many libraries that have fallen victim to a major disaster have taken as much as seven to ten years to restore.

Also, take care of yourself emotionally. Remember that you are a person too who has suffered trauma and now must face the greatest professional challenge of your life. Many times, managers are so busy managing the situation and caring for others that there is no time for their own needs. Just know that you will also experience a range of emotions so have some mercy on yourself. Forgive yourself for any mistakes. You are doing the best you can.

A disaster is a confusing time. Communication will be haphazard, changing and often inaccurate. Decisions will be made, countermanded, and made again. Buildings may be declared safe, then unsafe. All things are uncertain during this time. Just know that is the way it is. You must know and let the staff know that there will not be certainty about anything. Keep rolling with all of the changes and don't be frustrated with the ups and downs of disasters.

Importantly, maintain a positive attitude. Of course, it isn't easy in the face of sometimes heartbreaking disaster but it is vital that you believe, and transmit to the staff that you believe, that the library will be restored. It is essential to focus on the importance of the task ahead and the importance of library service in your community.

IN CONCLUSION

When coming back from major disaster, the first month is critical. Unfortunately, it is also the same time period when you and your staff are grappling with the first blow of the disaster. However, it is during that first month that the important groundwork is laid which will lead you and your staff to eventual success.

Nothing prepares us for a major disaster. Once a disaster strikes, we have these things at our command: our knowledge of libraries; our commitment to them; the library team; our brains; our courage; and our resourcefulness to bring back our libraries from disaster. The odd comfort is in knowing that many librarians have gone before us in fully restoring their libraries.

doi:10.1300/J118v25n03_03

The Seven Deadly Sins of Disaster Recovery

Randy Silverman

SUMMARY. Disaster response is largely about managing people. Human failings and vice can delay or derail library and archives recovery activities, causing irreparable damage to irreplaceable cultural property. Peter Waters' seven requirements for conducting successful disaster recoveries are contrasted with the Seven Deadly Sins to demonstrate the timeless nature of subversive factors at work in the disaster domain. Disaster planning is recommended as an indispensable alternative to running in circles. doi:10.1300/J118v25n03_04 *[Article copies available for a fee from The Haworth Document Delivery Service: 1-800-HAWORTH. E-mail address: <docdelivery@haworthpress.com> Website: <http://www.HaworthPress.com>.]*

KEYWORDS. Peter Waters, Florence flood, disaster plan, recovery protocol, library, archive

Randy Silverman, MLS, is Preservation Librarian, University of Utah Marriott Library, 295 South 1500 East, Salt Lake City, UT 84112 (E-mail: randy.silverman@utah.edu).

[Haworth co-indexing entry note]: "The Seven Deadly Sins of Disaster Recovery." Silverman, Randy. Co-published simultaneously in *Public Library Quarterly* (The Haworth Information Press, an imprint of The Haworth Press, Inc.) Vol. 25, No. 3/4, 2006, pp. 31-46; and: *Dealing with Natural Disasters in Libraries* (ed: William Miller and Rita M. Pellen) The Haworth Information Press, an imprint of The Haworth Press, Inc., 2006, pp. 31-46. Single or multiple copies of this article are available for a fee from The Haworth Document Delivery Service [1-800-HAWORTH, 9:00 a.m. - 5:00 p.m. (EST). E-mail address: docdelivery@haworthpress.com].

Available online at http://plq.haworthpress.com
doi:10.1300/J118v25n03_04

INTRODUCTION

Speed and effectiveness of the human response following a disaster are the most critical variables affecting the condition of an institution's collections in the aftermath of the recovery. Meeting the situation's shifting requirements calls for a well-conceived emergency response plan and people capable of competently implementing it. The loss of environmental controls and waterlogged collections that often accompany a disaster will continue to actively damage cultural materials until they can be stabilized. Time is of the essence; the pressures that ensue are often unique, complex, and destructive.

Disaster plans, when they exist–and only 20% of all U.S. collecting institutions have a written disaster plan[1]–are often drawn up by people with little or no firsthand experience in the recovery of water-soaked collections, and are likely to cause further problems upon implementation. And yet, effective disaster planning continues to be avoided–perhaps understandably. After all, planning for a disaster is essentially mapping a trip to hell, and who in good conscience wants to contemplate that adventure?

In practice, the unexpected is inherent in every disaster. They elicit variable human responses and are difficult to anticipate when drafting a disaster plan in the relative comfort of a dry office. Some people will rise to the occasion, others won't, and the suspension of normalcy can open up a subterranean experience not unlike the one in Lewis Carroll's *Alice in Wonderland.*

Mr. Carroll provides a glimpse of this world following the flood created by Alice's own tears when she was nine feet tall. Shrunken to a tiny height and drenched along with the other creatures, Alice and her companions readily acquiesced to the bedraggled Dodo's suggestion to convene a hastily concocted "Caucus Race" to dry out, though none of the party understood exactly what this entailed. The race, with its veiled reference to political campaigns, began by plotting a route "in a sort of a circle." Each of the participants was randomly "placed along the course, here and there." With no precise starting point, the racers simply "began running when they liked and left off when they liked." Under the Dodo's tutelage, a gentle, aerobic pandemonium ensued until "they had been running half an hour or so, and were quite dry." At this point, "the Dodo suddenly called out 'the race is over!'" The party came to rest. Panting breathlessly, each pressed the Dodo, quizzically asking, "but who has won?"[2]

Who indeed. A truth embodied in Mr. Carroll's frenetic tale is that the Dodo–a voice of authority in an unfamiliar situation–exerted control because the good natured troupe was caught off-guard, unprepared to act. While the immediacy of his response was laudable, was air drying by physical perambulation the most effective solution, even in Wonderland? Is running in circles after a disaster really an optimal response?

When dealing with historic, culturally significant, or potentially irreplaceable collections this question becomes critical, and the answer, it turns out, is largely dependent upon the capability and insights of the person directing the recovery. In 1995, Peter Waters, the father of modern book conservation in the United States and the recovery coordinator for the Biblioteca Nazionale immediately following the devastating Florence flood of 1966, defined seven essential requirements for successful recovery.[3] Having participated in numerous extremely large library and archive disasters throughout his career, Waters' observations remain poignantly insightful today.

In fact, it is telling to contrast Waters' seven characteristics for recovery success with the Seven Deadly Sins, that Fourth Century metaphor of human weaknesses[4] that describes an array of personal failings still undermining society today. More than a decade since Waters' remarks were first made public, unexpected problems continue to point to the unforeseen pitfalls that inhabit real disaster situations–chief among them, the individual shortcomings of the participants themselves.

SLOTH

Mr. Waters' first determinant for a successful recovery of cultural property, "**[t]he degree of human reaction and responsiveness under stressful conditions**,"[5] is easily undermined by Sloth, the sin of laziness.

People's ability to act decisively in stressful circumstances will largely decide the success or failure of a response, and institutional response rates can be delayed. For example, if directors and assistant directors are away from the office, out of town, or otherwise unreachable, the subordinate staff's hands are tied if they have not been pre-authorized to make decisions on their behalf. A variation on this problem is the normally dynamic director who becomes psychologically paralyzed by the immediate impact of the event and is incapable of making critical, rational decisions. The scenario of the "walking wounded" is far more common than one might suspect, and post traumatic stress (PTS) syndrome can incapacitate one person while leaving another unaffected even though both people witnessed the same event.

A well-conceived disaster response plan needs a 24/7 telephone tree and the ability to accommodate uncertainty by pre-establishing criteria for implementing alternative chains of command. Well before an emergency occurs, it is eminently desirable to identify and train a calm, well-organized person and his or her backup to take responsibility of the recovery process, and to delegate to them the authority needed to make expeditious decisions. This is especially useful in the case of an insightful director who knows he or she is far better suited to tasks such as talking with the media or addressing bureaucratic exigencies. It is also appropriate to establish a training regimen for everyone on staff, so, as with responses to fire alarms, institutional emergency response protocols become automatic.

Pre-selecting a trustworthy commercial disaster recovery firm and identifying one or more conservators experienced in disaster recovery to act as consultants significantly improves an institution's ability to move forward quickly. A conservation consultant will provide timely insights about which technical options are most appropriate, establish cost projections, and act as the institution's advocate in discussions with the commercial recovery firm and insurance adjusters (when insurance coverage is available), thus ensuring the collection receives appropriate treatment from stabilization until it is returned to the shelves. Commercial disaster recovery firms are capable of marshaling large amounts of labor quickly and are trained through experience to expeditiously implement recovery strategies. A timely response in a highly charged situation is critical, but, as Mr. Waters suggests, the leader's effectiveness and technical expertise will be key to his or her success.

If the response is neither timely nor effective, the institution's financial stability can be severely jeopardized long after the disaster, even in the nonprofit world of higher education. One year after the 1994 earthquake at California State University, Northridge, students–although initially tolerant about attending classes outdoors and, later, in temporary accommodations–began transferring to other universities to regain access to basic academic amenities such as classrooms, laboratories, and a library. This departure left California State Northridge increasingly under-funded, slowing the rate of building repairs, and initiating a downward economic spiral that ultimately threatened to bankrupt the entire university.[6] Learning from this example, Colorado State University, after its campus-wide flood in 1997, implemented a free daily shuttle service to transport students and faculty to and from other regional libraries. It also acquired the newly developed Ariel technology to provide electronic document delivery directly to departmental offices all over campus, significantly decreasing user dissatisfaction during the two years spent restoring the library's building and repairing 425,000 damaged books.[7]

ENVY

Mr. Waters' second crucial element for a successful disaster recovery, "**[t]he emergence of a leader who is able to handle the situation, can create order out of chaos and has the strength and courage to act decisively without fear of future recrimination,**"[8] may come to blows with Envy, the begrudging self-interest that seeks to illegitimately undermine a leader's success.

An effective disaster response coordinator may be the institution's director or someone within the institution's ranks who has the director's support. This person must be capable of making snap decisions carefully and conducting the recovery cost-effectively. The longevity of every damaged item in the collection hangs in the balance. To succeed, it is helpful if the disaster response coordinator has strong convictions, a sense of humor, and a tolerance for abuse because when the accusations begin, a resilient constitution is an indispensable shield.

Every step of the way, the leader's work can be supported or hampered by people within the very institution she or he is striving to serve. Conflicting priorities, while not always illicit, can also lead to contentious problems. Following the fire at the Royal Saskatchewan Museum in Regina, Canada in 1990,[9] conservators realized that handling soot-coated organic materials before they were cleaned caused the soot particles to become deeply embedded within the object's surface. Administrators, however, hastened to reopen the building to patrons. The smoke damaged collection was hurriedly moved offsite to accommodate building restoration with the result that some objects became far more difficult or impossible to clean. As a consequence, parts of the collection were ruined.

Another example includes the laying of traps to derail the career of Emanuale Cassamasima, Director of the Biblioteca Nazionale in Florence. In 1966, Cassamasima, a charismatic, pivotal leader and a scholar with a deep and abiding passion for books, was also a Communist engaged locally in the pursuit of social justice. At the time of the flood, his superiors in Rome were so antagonistic towards him because of his political persuasion that when he asked for emergency help they simply failed to respond to his communications. Rather than risk the total loss of the library's collection by delaying, Cassamasima diverted his acquisitions budget to pay for the immediate transport and drying of the library's wet books.

Lacking explicit permission from his superiors in Rome, this act amounted to insubordination, threatening Cassamasima's ongoing employment. Immediate international intervention in the form of American, Australian, and British aid buoyed him up to persist in the work of building a restoration lab that ultimately employed over 100 Italians trained to the highest conservation standards by Peter Waters, Tony Cains, Chris Clarkson, and Don Etherington, among others. Over the next three years, ongoing foreign charity permitted Cassamasima to persevere until his supervisors conceded. Their attempted retribution through the unethical exercise of authority was narrowly averted because the events in Florence were played out on a world stage. Cassamasima's moral conviction is all that prevented the total loss of nearly 1.5 million books, at least 90,000 of them rare.[10]

GREED

Mr. Waters' third requirement for a successful recovery concerns "**[t]he actual methods used to determine an assessment of loss and damage leading to a plan of action which ultimately affects the costs of reclamation, replacement and restoration.**"[11] This assessment is easily skewed by self interest, Greed, or avarice once the rules of normal commerce, including competition, have temporarily been suspended.

The most egregious example of this occurs when disreputable recovery companies inflate prices, insist on providing nonessential services, or produce poor quality workmanship to profit from the confusion that dominates the disaster domain.

This predatory environment can be modified dramatically by carefully pre-selecting a disaster response company long in advance of the event. By initiating a formal Request for Proposal (RFP) process and inviting reputable companies to participate,[12] current price ranges for comparable services can be established. While past pricing cannot be applied directly to future market conditions, nor the specific scope of work needed be known in advance, the RFP process offers institutions insight into the basic cost structure of disaster recovery services. This evaluation must go beyond fiscal considerations, however, and probe the quality of performance received from the client's perspective, an issue that will require a bit of sleuthing. The RFP can contain a request for disclosure about past customer satisfaction, with the accuracy of these statements substantiated through direct contact with the clients. However, untrustworthy companies are not likely to list dissatisfied customers among their references. Additional information about a firm's past performance can be gleaned through personal conversations with seasoned conservation consultants able to provide direct observations from the field or indiscretions they are aware of but are not at liberty to publish.

When one or more insurance companies are involved in a recovery, it is in their best interest to keep recovery costs as low as possible. This fiscal conservatism often includes adjusters applying pressure to encourage early settlement, ostensibly to expedite the recovery process but in point of fact to "fix" recovery limits before unexpected factors crop up to inflate the bottom line.

In the first days following the Colorado State University library flood, the institution's emphasis was on packing out and freezing the collection to stabilize it as quickly as possible. It was difficult then to anticipate the degree to which mold would become a factor. But, the 14 days required to transport 425,000 sodden books to an out-of-state commercial freezer plant caused rampant mold growth. It took months of subsequent investigation to determine the long-term human health risks posed by such a huge concentration of mold, whether sterilization was required, and if so, which technique was most appropriate (information sorely lacking in the library disaster recovery literature). Not surprisingly, the primary insurer tried to settle the claim as early as the third day into the recovery, long before the additional cost of sterilization was even a remote possibility.

This sort of problem also crops up when a disaster recovery contractor negotiates a bonus for completing the restoration by a specified date. Again, at Colorado State University, the administration interpreted the inclusion of this contractual clause as meeting their need to quickly normalize university operations and, after much debate, agreed to the stipulation. In practice, however, this contingency provided the contractor with an incentive to cut corners. After three months, this author, as conservation consultant, called for a reevaluation of the protocol for dealing with the library's mold problem, as the university's exposure seemed palpable. In turn, the university called for an independent assessment to ascertain whether vacuum freeze drying and rebinding without sterilization was adequate or whether sterilization or replacement of the library's 20th century collection was a more appropriate option. Peter Waters and Norbert Baer were commissioned to provide this second opinion and their conclusion corroborated this consultant's concerns. However, both consultants' concerns were summarily ignored by the contractor,[13] who continued implementing his original recovery protocol. It was only through the intervention of microbiologist Doug Rice and recovery specialist Larry Wood that an economical sterilization protocol using gamma radiation was developed, and the university overruled the contractor's objections.[14]

Monitoring the actions of this type of disaster contractor is in the best interest of both the institution and the insurance company involved in the loss, but even the people hired to provide this type of oversight must be closely scrutinized. Independent disaster contractors comprise a very small group; when one is hired by an insurance company to look out for its fiscal interests, they are essentially being paid to scrutinize the work of a close professional peer. In practice, these personal ties create more than a casual conflict of interest. The overseer has reason to support a colleague's decisions because on subsequent jobs, these two may well reverse their roles.

GLUTTONY

Mr. Waters' fourth criterion for success following a disaster is "**[t]he degree to which management can develop strategies to cope with untimely, insistent and inaccurate media reporting.**"[15] The glut of media attention brought on by the disaster can reinforce or seriously undermine an institution's credibility. Causes of misinformation include: well-meaning but ill-informed staff who inaccurately report the cause or extent of damage; reporters who rely on popular rumor rather than fact; or appropriate authorities who do not thoroughly disclose the institution's situation in a timely fashion.

An institution's media releases must be accurate, truthful, and to the point. Responsibility for these reports should rest with one key person within the organization. No matter how grim the news, public support can be rekindled by timely, consistent updates that can generate support, both during and after the crisis.

ANGER

Fifth among Mr. Waters' requirements for a positive outcome is the leader's "**[l]evel of determination and strength of character to continue the recovery process while being confronted by negative criticism from those who seek to find someone to blame for the catastrophe.**"[16] To be effective, the disaster recovery coordinator must exude personal integrity and discharge his or her duties dispassionately with a clear absence of anger.

Identifying the root cause of a disaster may seem like a step toward preventing similar events from reoccurring in the future, but ultimately solving the immediate crisis can't be accomplished through blame or anger. In the 2002 Prague flood, governmental authorities announced that the Vltava River's impending rise would be relatively insignificant. Placated by this information, museum administrators took no steps to safeguard or relocate their collections. Three days later their holdings were submerged in Prague's most destructive flood of the millennium. Describing these events a year later, museum personnel remained angry beyond words, betrayed by the incompetence of politicians who, in an effort to avoid panic, chose to suppress early warnings available from the region's river monitoring technology.[17] Energy focused on this misleading public announcement, however, did nothing to rectify the still devastated situation.

A conservation consultant hired to advise on recovery protocols is not immune from similar frustrations pouring out from all affected parties. His or her character will be sorely tested by people who may inadvertently or intentionally create obstacles to undermine the recovery effort. Misdirected anger from a single influential person can rapidly turn the tide of opinion against a consultant and reduce his or her status to that of an interloper, or even a scapegoat. While this behavior may seem irrational, it should be remembered that vilifying the stranger is a common human trait. Good Samaritan laws in the U.S. were drafted, after all, to protect the well-intentioned efforts of those providing emergency medical assistance to people regardless of the success of the action or the victim's feeling about the competence of the intervention.

The conservation consultant must remain vigilant and confront attempts aimed at subverting the recovery effort. Inevitably, the collection will be scarred by the event and all parties need to acknowledge from the outset that their hard work is going to produce a less-than-ideal outcome. Disasters occur within a preexisting social and political environment that may prove as complex to comprehend as the technical challenges required to stabilize, dry, and repair damaged collections. The consultant's recommendations must remain consistent and credible if they are to be effective.

In the aftermath of Hurricane Katrina, this author joined the efforts of HEART, the Heritage Emergency Assistance Recovery Teams organized by the American Association for State and Local History and the American Institute for Conservation. On 26 September, 2005 we visited the Old Spanish Fort Museum in Pascagoula, Mississippi. This one-room local museum had been inundated by approximately two feet of standing water, and had not been opened up in the six weeks following the storm. An oozy layer of mud covered the museum's floor, and both standing water and collections remained inside locked exhibit cases, the surface of most organic objects thick with five or six varieties of active mold nurtured by the lack of air movement. Despite swollen wooden window frames, the HEART team pried open the tiny museum's six double-hung windows to create a bit of cross ventilation and informed the elderly curator and his volunteer assistant that their artifacts needed outside airflow to dry or they would continue to mold. We then moved on to assess other damaged institutions.

Visiting the Old Spanish Fort Museum one week later, the second HEART team discovered tightly closed doors and windows, re-sealed by the elderly curator and his assistant concerned about vandalism despite the museum's rural location. Again, the building was opened up to encourage air movement, but when the third team arrived the following week the museum was once more hermetically sealed with the mold bloom frenetically engulfing the collection. This obstinance could have driven the team to a display of anger had the situation not been so pitiable.

Remaining sympathetic is key to relating to the trauma people are experiencing at a personal level. Not all attempts to assist can succeed, but ideally, a mutually productive meeting of the minds can improve the situation's evolving exigency. Above all, everyone involved should strive to remain positive, supportive, and flexible to take advantage of developing opportunities.

LUST

Mr. Waters' sixth universal condition affecting the recovery's success, "**[t]he methods used to prepare and dry material which directly and irreversibly affect post recovery costs,**"[18] is a consideration easily linked with the Lust for money.

Some commercial disaster recovery firms routinely overcharge for services in addition to selling customers treatment options than are actually unnecessary. Being unfamiliar with the protocols necessary to address their collection's needs, most disaster victims are implicitly vulnerable to accepting inflated prices from unscrupulous contractors. After the flood occurred at Colorado State University, the contractor listed in the library's disaster plan was immediately hired, with no agreement on pricing in place. On the third day of the recovery, the head of the firm set a figure of $1.5 million to pack out the library–that is, to box and transport the wet books to a commercial freezer plant. This disclosure led to an emergency meeting with the university's president and his cabinet where this consultant estimated the pack out to be worth approximately one fourth that amount. In an unprecedented course correction, the president fired the first firm and, better informed, appointed an alternate to immediately take over the entire job.

Most disaster victims, however, rely on their contractor's expertise for guidance, leaving them vulnerable to being misled or fiscally devastated. Unscrupulous business practices are a product of the carnivorous atmosphere that can permeate the post-disaster environment where relationships are short-lived. Unsettled conditions accompanying the recovery period make it easy for people who are so inclined to prey on the disadvantaged through larcenous business dealings.

Well-intentioned but badly conceived treatment protocols can also drive up recovery costs and cause significant and irreversible collection damage. Following the 1966 Florence flood, when freezers were unavailable as a means of fending off mold growth, mud-laden books were immediately dried in commercial brick and tobacco curing kilns before conservators had a chance to intervene. A seemingly expeditious solution to the immediate problem, thermal drying caused the paper's gelatin sizing and the book's animal glue spine lining adhesives to migrate to the surface of each page, leaving many books baked dark brown and brittle. This outcome required that each book be taken apart section-by-section and washed to remove the scorched gelatin, a treatment regimen that dramatically increased recovery costs. Today, thermal drying is also known to damage cellulose, reducing paper strength by approximately 15%.[19]

Moreover, if sterilization is required, the choice of technique is significant in terms of paper permanence. For example, mold on Colorado State University's 425,000 water-damaged books was sterilized using gamma radiation in 1997, an experimental technique for cultural property. Although economically viable and efficient, gamma radiation in doses large enough to destroy mold spores is now recognized to cause polymer scission, reducing cellulose strength by approximately 25%.[20] This irreversible condition ultimately reduced the permanence of the material treated, underscoring the importance of choosing recovery methods that are technically appropriate for the job at hand.

PRIDE

Finally, Mr. Waters' seventh measure of disaster recovery success, "**[t]he willingness of an administration to openly share a recovery experience with others who are interested in developing a disaster preparedness plan,**"[21] is an act of humility that stands in stark contrast to the Pride preventing full and open disclosure of mistakes that negatively impact a recovery. If made public, lessons learned–however embarrassing–can inevitably help others avoid similar pitfalls.

In 1986, my first disaster recovery was initiated by a slow leak in a wet-pipe fire suppression system at the LDS Church's Family History Library that had gone undetected for weeks because it occurred above a compact shelving unit in a little-used part of the collection. I was part of a group of responding conservators who performed a "textbook" recovery. Several thousand books were rapidly packed spine down into plastic milk crates and moved via a human chain to vans waiting at the library's loading dock where they were whisked off for freezing in a local Salt Lake City ice plant. The recovery went like clockwork. When the team left that afternoon, every person was extremely elated by what had been successfully accomplished in a matter of hours. Two weeks later, however, we learned that the Hygea Ice Plant, the facility where the books had been taken for freezing, had burned! This eventuality is one I have never seen discussed in the disaster recovery literature and it provides an extremely sobering lesson about the capriciousness of life and the universal constant–that "anything that can go wrong, will" (Murphy's Law in the U.S.; Sod's Law in the U.K.)–even after a disaster.

Another example was shared by a Czech conservator following the 2002 Prague flood. The basement of the Pinkas Synagogue, one of the complex of buildings comprising the Jewish Museum in Prague's historic Jewish quarter, filled with water. The first reaction of museum officials was to bail out the synagogue's basement, which was summarily accomplished. What appeared to be a straight forward decision turned out to be ill-conceived, however, a fact that was not clearly understood until after the work was completed. Because of the vast reach of the flood, water pressure throughout Prague was impacted. The decision to remove the water from the synagogue's basement did not take into account the source of this water, which originated from a spring fed, fresh water well that constantly replenished the synagogue's mikva (a ceremonial bath for women). Once this clean water was removed, pressure from tons of standing water surrounding the building forced sewage to back up into the synagogue's basement, displacing the fresh water. The hubris of pride–assuming the full extent of a complex problem can easily be understood–actually contributed to making a bad situation far worse.[22]

Fear of recrimination or public humiliation often prevents institutions from sharing these types of miscalculations with colleagues. Honest, self-critical disclosure that identifies problems contributing to the disaster or inadequacies leading to a flawed recovery can help institutions reach closure in terms of resolving the disquieting consequences of a catastrophic event.

Equally important, other cultural repositories can learn from these hard-won lessons if we take the opportunity to help raise professional awareness. It is also important to note that pride can come before a fall and may lead to an institution's willingness to reject offers of assistance in times of emergency–to "go it alone." Stories of whole collections being discarded after a flood or a fire are disconcerting but are not uncommon; reaching out for help should be an institution's first response to a disaster affecting its collection.

CONCLUSION

No form of collection recovery can ever reverse the consequences of catastrophic damage. Thus, the best form of disaster response is prevention. Vital to preventing or mitigating potential devastation, all collecting institutions are well advised to invest the time and energy necessary to draft, implement, and update a disaster plan. This includes attention to the diverse guises of human weakness that can subvert an institution's best laid plans. Individual or corporate transgressions and misconduct represented by the classic seven deadly sins–Sloth, Envy, Greed, Gluttony, Anger, Lust, and Pride–are all likely to surface during an event large enough to disrupt normal patterns of commerce and social interaction.

Ultimately, a collecting institution's success in responding to a disaster will hinge on one simple precept: the speed and effectiveness of the reaction can decide the fate of the collections. Irreplaceable cultural property must be stabilized as quickly as possible as wet organic material will soon mold, some inks and dyes will run, and partially dried coated paper and photographic emulsion can block together. The efficacy of the recovery can be significantly improved if certain facts are understood *a priori*. If the collection contains irreplaceable artistic, historic, or cultural works, precautions should be implemented *beforehand* to properly house (box) valuable or fragile material to minimize potential water damage and prevent breakage. As possible, these objects should be stored above the high water line (the 500-year flood plain). A network of conservation consultants should be established in advance to ensure the institution can draw upon technical expertise immediately as needed. And, these consultants need to be experienced in handling large-scale disasters and highly skilled in material-specific drying techniques pertinent to the institution's holdings.

Finally, like all other work-related activities, disaster response turns out to be largely about managing people. Efforts to familiarize institutional staff with disaster recovery protocols and to involve authorities from the university, city, county, and state (including the local fire department) in the planning process will pay off should a real disaster strike. Pre-selecting a reputable professional disaster recovery firm and authorizing several people within the institution to contact it if its services are ever required is a proactive and responsible step. Making agreements in advance with local and regional vendors can also streamline the process of procuring requisite materials and services such as paperboard boxes, plastic garbage can liners, transport vehicles, and freezer space.

Disaster planning may indeed be mapping a trip to hell, but heaven help the organization that has to make that journey bereft of a plan. While Peter Waters' seven basic prerequisites for success in the recovery field seem simple enough, in practice, they are subject to being undermined by human frailty and vice, as he was well aware. Investing in the disaster planning process is one way institutions can preempt potential problems–for the dubious, however, running in circles may still suffice.

NOTES

1. Heritage Preservation and the Institute of Museum and Library Services, *A public trust at risk: the heritage health index report on the state of America's collections* (Washington, DC: Heritage Preservation, 2005): 6-7.

2. Lewis Carroll, *Alice in Wonderland*, illustrated by Maraja (New York: Grosset and Dunlap, c. 1963): 23.

3. Peter Waters, "From Florence to St. Petersburg: An Enlightening and Thought-provoking Experience," (19 August 1995), paper written for the conference, Redefining Disasters: A Decade of Counter Disaster Planning, held at the Library of New South Wales, Sydney, Australia, 20-22 September 1995, unpublished typescript: 7.

4. First assembled as eight sins in the fourth century by Evagrius of Pontus (d. c. 400); see Michael Bloomfield, *The Seven Deadly Sins* (Lansing: Michigan State College Press, 1952): 59-60; cited in Robert C. Solomon (ed.), *Wicked Pleasures: Meditations on the Seven "Deadly" Sins* (London: Rowman & Littlefield Publishers, Inc., 1999): 47, n. 5.). Pope Gregory the First, also known as Gregory the Great (d. 604), instituted the current list of seven during the fifth century; see Solomon: vii, 17, 21, 33, 54, 55.

5. Peter Waters, "From Florence to St. Petersburg,"1995: 7.

6. *Academic Aftershocks: January 17, 1995, One Year after the Northridge Earthquake: California State University, Northridge*, Videorecording (47 min.): sd., col.; 1/2 in. (Northridge, California: California State University, Northridge, Design Media, Inc., 1995).

7. Camila A. Alire, *Library Disaster Planning and Recovery Handbook* (New York: Neal-Schuman, 2000).

8. Peter Waters, "From Florence to St. Petersburg," 1995: 7.

9. Sarah Spafford-Ricci and Fiona Graham, "The Fire at the Royal Saskatchewan Museum, Part 1: Salvage, Initial Response, and the Implications for Disaster Planning," and, Part 2: Removal of Soot from Artifacts and Recovery of the Building," *Journal of the American Institute for Conservation* 39, no 1, (Spring, 2000), p. 15-36 and 36-56.

10. Peter Waters, "From Florence to Leningrad: A Learning Experience," paper written for the U.S.S.R Library of the Academy of Sciences (BAN) conference, 1995, unpublished typescript: 4; and, Peter Waters, "Book Restoration After the Florence Floods," in Herbert Spencer (ed.), *Penrose Annual* 62 (1969): 83.

11. Peter Waters, "From Florence to St. Petersburg," 1995: 7.

12. This author is happy to share with interested parties an RFP developed by the University of Utah to pre-approve a commercial disaster recovery firm, the contract that resulted from the RFP process, and a list of reputable disaster recovery firms. <randy.silverman@utah.edu>

13. Personal communication with Peter Waters, 14 March 1998.

14. Randy Silverman, "The Day the University Changed," *Idaho Librarian* 55, no. 3 (February 2004) <http://www.idaholibraries.org/newidaholibrarian/200402/index.htm> or <http://content.lib.utah.edu/u?/ir-main,103>.

15. Peter Waters, "From Florence to St. Petersburg," 1995: 7.

16. Peter Waters, "From Florence to St. Petersburg," 1995: 7.

17. Personal communication with colleagues attending the Czech and Slovak Library Information Network (CASLIN) conference, "Crisis Management and Recovery," Holenský Dvur, Czech Republic, 8-12 June 2003.

18. Peter Waters, "From Florence to St. Petersburg," 1995: 7.

19. Unpublished results from research comparing contemporary mass paper drying and sterilization technologies, supported by a generous grant from National Park Services, National Center for Preservation Technology and Training, and conducted in

conjunction with the British Library and the National Library of the Czech Republic, 2004-2006.

20. Ibid.

21. Peter Waters, "From Florence to St. Petersburg," 1995: 7.

22. Personal communication with Jerzy Stankiewicz, Head of the Restoration Department, National Library of the Czech Republic, Prague 12 June 2003.

doi:10.1300/J118v25n03_04

The Importance of Disaster Planning for the Small Public Library

Sonya L. Green
Thomas H. Teper

SUMMARY. Most articles about disaster planning in libraries focus on large institutions such as major academic, national, and public libraries. There has not been much focus on the importance of disaster planning for smaller public libraries. This article aims to help fill in the gap using a combination of case studies, data from the 2005 Illinois Statewide Preservation Needs Assessment, and existing disaster planning documentation. Education and training, awareness of resources, and creating a disaster plan are the points stressed as the most important factors a small library can change to affect the safety of its staff, building, and collection. doi:10.1300/J118v25n03_05 *[Article copies available for a fee from The Haworth Document Delivery Service: 1-800-HAWORTH. E-mail address: <docdelivery@haworthpress.com> Website: <http://www.HaworthPress.com> © 2006 by The Haworth Press, Inc. All rights reserved.]*

KEYWORDS. Disaster planning, public library/libraries, preservation, conservation, grants, education and training

---

Sonya L. Green, MLS, is Youth Services Librarian, Millicent Library, 45 Center St., Fairhaven, MA 02719 (E-mail: sundaykofax@gmail.com).

Thomas H. Teper, MLIS, is Head of Preservation, University of Illinois at Urbana-Champaign, 246F Main Library, 1408 W. Gregory Dr., Urbana, IL 61801 (E-mail: tteper@uiuc.edu).

[Haworth co-indexing entry note]: "The Importance of Disaster Planning for the Small Public Library." Green, Sonya L. and Thomas H. Teper. Co-published simultaneously in *Public Library Quarterly* (The Haworth Information Press, an imprint of The Haworth Press, Inc.) Vol. 25, No. 3/4, 2006, pp. 47-59; and: *Dealing with Natural Disasters in Libraries* (ed: William Miller and Rita M. Pellen) The Haworth Information Press, an imprint of The Haworth Press, Inc., 2006, pp. 47-59. Single or multiple copies of this article are available for a fee from The Haworth Document Delivery Service [1-800-HAWORTH, 9:00 a.m. - 5:00 p.m. (EST). E-mail address: docdelivery@haworthpress.com].

Available online at http://plq.haworthpress.com

doi:10.1300/J118v25n03_05

INTRODUCTION

The protection of public libraries and historical societies took on a new urgency in the aftermath of Hurricane Katrina. Following on the heels of coverage devoted to civil and charitable organizations' efforts to rescue the many stranded people, newspaper and television coverage of the devastation seemed unrelenting. Amidst this torrent were many images of devastated libraries. The images of waterlogged books, collapsed ceilings, pervasive mold, and tidelines marking the height of the advancing storm surge elicited a sense of hopelessness, sympathy, and a haunting form of motivation for many librarians that never faced the type of disaster faced by our colleagues in the Gulf Coast. Despite all that libraries have heard about protecting collections from disasters, how do we prepare our institutions to face such cataclysmic disasters? And, why should we, given the seemingly unstoppable nature of such a crushing event?

This sense of impending doom among libraries is not new. Thirty years ago, images of massive flooding along Italy's Arno River Valley in 1967 mobilized librarians and conservators from across Europe and the United States to assist in recovering millions of damaged books and manuscripts. Focusing on the devastated city of Florence, this effort received feature coverage in the July 1967 *National Geographic.* Preservation administrators and conservators consider this recovery effort a watershed moment in the preservation of library and archival materials, bringing world-wide attention to the needs of our shared cultural resources.

In the United States, a massive fire at the Los Angeles Public Library rekindled the sense of purpose imparted by the Florence Flood two decades earlier. Actually consisting of a series of three fires, the first two set by an arsonist and the third kindled in construction debris during the rebuilding, the arson fires damaged and destroyed hundreds of thousands of volumes. Television coverage of the fires captivated the public and lent substantial support to the still emerging field of preservation. First responders standing amid the rubble declaring that one or two sprinkler heads could have contained the fire and hundreds of volunteers sifting through the damaged collections for salvageable material emphasized the benefits of both prevention and preparedness activities.

While massive floods, fires, and storms drew attention to the needs of disaster preparedness and mitigation, they only represent part of the tale. The pages of *American Libraries* and local newspapers are all too often populated with articles and news briefs detailing accidental and purposely set fires, water main breaks, mold blooms, and storm damage to public libraries that never receive wider professional or media attention. While disasters are often unavoidable, those that hit smaller public institutions rarely receive the attention they deserve.

Local libraries play a key role within their communities, serving as community centers, hubs for face-to-face and virtual networking, and the primary access point for many individuals seeking access to a wider world. Local libraries frequently serve as the primary access point for individuals seeking access to their local community's history. Holding special collections of significant local import, public libraries serve the curious, professional historians, and genealogists, providing them with insight into local economics, demographics, politics, history, and culture. Preserving these resources in the face of a disaster presents a significant challenge for any institution, but protecting locally valuable collections remains difficult in light of competing interests. What can the public librarians do to help plan for such an event?

DISASTER PLANNING FOR SMALL PUBLIC LIBRARIES: A REVIEW OF THE LITERATURE

Although there is a good deal of scholarly work written about the preservation and conservation of libraries, the vast majority of works are aimed at large library systems with financial and staff resources available to conduct a large-scale response. While the general information in these pieces can be applied to smaller libraries, what seems to be missing is the focus on what smaller libraries should be doing to preserve, and in the event of a disaster, conserve their collections.

There are articles about disasters for small libraries available; however, they are few and far between. Much of the disaster literature is event-based, focusing on descriptive accounts of a particular disaster in one particular library. These articles can be helpful if the information can be applied to one's own library, but the goal of this article is to provide guidance and best practices for all small libraries.

Being prepared remains one of the few things a library can do in the event of a disaster. There are a number of articles that outline the guidelines for the disaster planning process. Often aimed at large libraries, some of the guidelines are outside of the scope of a small library, but the process of decision-making is universal. Bolger (2003) is particularly useful because the guidelines are written in general terms (assign responsibility, set up liaisons with local emergency agencies, assess potential hazards) and described in a general way.

Within the topic of disaster planning is the preparation of disaster kits. Many of the lists of requisite supplies are too intricate for all but the largest institutions. Large public libraries are expected to have a storage area with generators and water pumping equipment, whereas a small library might have room for a box in a closet with the minimum tools.

While size may impact the types of equipment available, a library's size alone should not adversely impact its level of preparedness and ability to respond to a disaster. The preparedness might be more creative, priority setting might take different directions, and response activities may not be wholly an internal affair. Reliance upon community resources may be the most valid way of providing protection against the threat of disaster. Much of this information can be found in the article "The Effect of Disasters on the Rural Library" by Ellen Ashbauh (2003). Ashbauh's article provides a good starting point for small or rural libraries, covering the gamut of disaster planning and response topics–planning for a disaster, and the specific challenges that small and rural libraries experience after a disaster.

Developing an up-to-date disaster plan is one of the few things that any library can do to help mitigate the eventual impact of a disaster. There is a great deal of information available in books and journals about how to write a plan, but these are usually aimed at large libraries. There are boilerplate plans available through regional organizations like the Northeast Document Conservation Center (NEDCC, 2006) that can be pared down for a small library. "You Too Can Do a Disaster Plan" (Kane, 2001) features information about creating a library-specific disaster plan from scratch, with no prior experience.

The key to disaster preparedness in a small public library is to examine the needs of the collection and establish priorities for salvaging those collections most necessary to the library's operations and most important to the local community. While most of a small public library's circulating collection may be replaceable, the items that are of local importance are often one-of-a-kind.

THE IMPORTANCE OF LOCAL CONCERNS IN LOCAL COLLECTIONS

The value of local history collections is growing. The democratization of genealogical research, the increased interest in what are increasingly viewed as valuable historical artifacts, and the spread of popular print and televised histories have all contributed to this situation. Likewise, while many librarians and archivists initially speculated that the availability of digitized resources would reduce the need for on-site access, the opposite appears to be true–at least where special collections are concerned. Consequently, local collections are receiving more attention than they have in decades–both in terms of use and in terms of financial support.

While funding sources are available to support the preservation of local history collections, the prioritization and attention devoted to these collections is still largely a matter of local concern. Built around kernels of local resources, these collections are often rich, providing a level of personal insight that published histories cannot parallel, and their preservation remains primarily a local concern. Built, managed, served, and most highly valued locally, these collections are local treasures. While many organizations and individuals can provide professional advice on their care, the special collections in local libraries and archives are local treasures, requiring local input for their development, documentation, and preservation.

Among research libraries, one of the greatest challenges in disaster preparedness is the development of priorities. The thought of a catastrophic disaster descending upon a collection that librarians have spent decades constructing and managing is almost too painful to contemplate. The possibility that one part of the collection–representing many lives' work made tangible in bound form–could be ranked a higher priority than others is unbearable. However, the presence of established priorities is among the most important factors in an institution's recovery from a disaster.

For those involved in developing and managing collections at public libraries, there are some important questions to ask. If the local fire company successfully ensures the safety of patrons and staff and contains a fire, does the local fire company know what collections are the most important to protect? Do the librarians and staff within your institution know which collections are most important to protect? In the days following a disaster, will the most valuable collections be salvaged first? Do the librarians and staff at the institution know which collections are the most valuable and can they clearly instruct volunteers and commercial recovery firms how to find these materials? These are questions that local administrators must ask themselves because knowing the historic and intrinsic value of collections is a local concern.

A SURVEY OF DISASTER PREPAREDNESS AT THE LOCAL LEVEL

Local leadership and coordination remains crucially important in disaster preparedness and mitigation activities. However, the reality is that few institutions actively work to prepare their institutions. Nationally, the level of preparedness for disasters remains critically low. Initiated in 2003, the Heritage Health Index sought to evaluate the level of preservation activity and preparedness among cultural heritage institutions. Completed by the non-profit organization Heritage Preservation with the support of the Institute of Museum and Library Services, the Heritage Health Index reported that 78% of all responding libraries–including public, academic, and special–possessed no written disaster plans (Heritage Preservation, 2005).

The figures may be better on a local level, but they are hardly reassuring. Completed in 2005, the *Illinois Statewide Preservation Survey Project* evaluated public, academic, and special libraries, historical societies, museums, and other cultural heritage institutions. Although 27% of public libraries indicated training in disaster preparedness and recovery to be a priority and 19% of the total respondents indicated suffering from a disaster that damaged materials within the last five years, disaster preparedness and mitigation activities among the respondents remain low. Despite the lessons presented by the L.A. Public Library Fire, only 86% of respondents possess fire extinguishers in all of some of their facilities. More advanced fire protection such as smoke detectors, alarms, and sprinklers are only found in 75%, 71%, and 33% of responding institutions. Perhaps even more alarming, the lowest cost option available to any institution, the preparation of a disaster plan, was only reported as completed in 52% of the responding public libraries (Clareson et al., 2005). While better than national figures reported in the Heritage Health Index, this is not reassuring.

PROTECTING LOCALLY IMPORTANT COLLECTIONS FROM DISASTERS

Local collections can consist of local newspaper archives, ephemera, diaries, scrapbooks, books by local authors, and any other original or one-of-a-kind materials that are of or about the community. These can include regional materials as well. Although these materials are not necessarily financially valuable, the loss of such artifacts to a local community could be devastating. Protecting these items ensures the protection of local historical research and scholarship.

Planning and Prioritizing

The first step for protecting local collections from disasters is planning. Forethought may mean the difference between saving and losing the collection, and the time and financial resources spent creating a useable plan can save an institution from significant additional financial or emotional loss if a disaster is averted or its impact is minimized. This goes for the library collection as a whole, but applies most specifically to locally important items. Part of the disaster plan should be a prioritization of materials. In most public library collections, large portions of the collection are either currently in print, available in a reprint, or less important to the library's primary constituents. Much harder to repurchase are the local items. In the event of a disaster, a library's local collection may be the library's highest priority and should be specifically named within the disaster plan. Knowing where the local collections are and how to evacuate them is crucial to their salvage.

A disaster does not have to be a ravaging fire or flood. A small leak or high humidity could cause a mold bloom to occur in the basement of the library where the old city records are kept, and if no one looks in on them, they could slowly be destroyed. Assessing the storage locations and housing materials of special collections may prevent a disaster due to neglect.

Part of disaster planning involves developing a decision tree that outlines salvage requirements. The decision tree for salvageable materials will differ from that of the general collections. Judith Fortson's *Disaster Planning and Recovery: A How-To-Do-It Manual for Librarians and Archivists* provides examples of such decision trees. As noted previously, the "financial value is minimal, but the inherent value to the community is immense" (Matthews and Feather, 2003 p. 194). Replacing a local collection is difficult, both because of the uniqueness of the items, such as the diary of a local resident, and the expense to replace them. If an item is available, it might be through a rare book dealer or auction. The decision might be to discard waterlogged materials from the popular collection, but salvage a one-of-a-kind handwritten novel, even if it is charred, wet, and muddy. But, without developing a decision tree that can be used by all members of the Library potentially associated with coordinating a recovery effort, significant resources can be spent trying to salvage less important portions of the collection.

Inventorying

Part of being disaster-prepared means knowing what items you have, in case they are destroyed, missing, or left unrecognizable. Often, the circulating collection is in the online catalog, but the local collection is not. If the only set of records is with the collection itself, a library could be left with no idea of what has been recovered. Keeping an off-site copy or digital copy of the inventory provides a level of security even if resources are not available provide full cataloging of the collection (Matthews and Feather, 2003).

PLANNING FOR THE LIBRARY'S RESPONSE AND RECOVERY EFFORTS

Communicating About Priorities and Plans

Beyond developing a disaster plan specific to an institution's needs, the most significant component of disaster planning focuses on communication. This communication largely entails building contacts with local disaster-response vendors and first responders and training personnel. Commercial firms and first responders will be much more valuable if they are familiar with the library and the disaster plan. In the case of first responders, they frequently appreciate being invited into facilities to tour, ask questions, and communicate with personnel about risks and challenges in responding to a disaster. When responding to a fire, the role of the fire department is to protect lives, contain the fire, and then protect property. Often, these activities involve great personal risk to the responding fire company, and information learned in the visits can be beneficial to both parties in helping to protect individual lives.

Staff should be familiar with the disaster plan as well. Walking through the plan and talking through a disaster scenario will familiarize staff with their roles and will make responding to a real emergency less stressful. Copies of the disaster plan should be kept outside the library. Matthews and Feather (2003) suggest keeping copies at the homes or in the cars of key personnel. Staff may feel overwhelmed after a disaster, which makes having a clear plan to follow all the more important.

As discussed in Wellheiser and Scott's *An Ounce of Prevention* (Wellheiser and Scott, 2002) a successful recovery effort may include the support of the local community. Public libraries often have very strong and dedicated supporters in their town, and the power of the community may be an institution's best resource. Volunteers can search for local books on eBay, clean up debris, and donate books to help get the library back up and running. This increases community involvement with the library, which will have positive effect upon the library staff post-disaster, and create a common goal for the community as a whole to participate.

Rebuilding the Collections

In the event that portions of local collections are lost, there are ways to rebuild the collection. Matthews and Feather (2003) suggest asking the community for support. Asking for donations could be a fast and cheap way to rebuild the collection. Making available the local collection record list of damaged materials might yield donations. Another option is to ask for loans of local materials to be digitized or reproduced as facsimile copies for patron use. This would require the equipment and staff hours to scan or funds to contract for services. If no other options were available, scanning would mean that the material is accessible to the public again, and it would have the added benefit of being digitized. However, given other needs within the organization, this might not be a priority activity.

Local collections are only a fraction of the average public library's collection. In considering the replacement of the general collections, keep in mind that each and every title does not require replacement. In some cases, outdated or less used materials can be weeded as part of the salvage process–discarded before they are even dried. However, this should be undertaken in as thoughtful a manner as possible. Depending on the severity of the disaster, replacing the popular collection might require developing a multi-tier plan–identifying priority collections for replacement. If the library is running a temporary service until full operations are resumed, acquiring more lending-friendly replacements and fewer academic titles would be more appropriate, and the temporary collection could be absorbed into the permanent collection once the building is ready (Matthews and Feather, 2003).

As discussed in the context of local collections, donations are one way of repopulating the shelves. The books donated may not exactly fit with the library's eventual needs, but it is appropriate to use donated books in a book sale that would generate funds for subsequent purchases. However, donations require evaluation, so although the cost for added titles remains low, the storage and staff hours required for processing must be considered.

There are several tactics for repurchasing materials. Matthews and Feather (2003) outline a "divide and conquer" plan, where staff split up the call number range and purchase for a designated section, which minimizes duplication. Another idea is to enlist the help of staff at nearby libraries to help choose materials. This has an increased chance of duplication, but with communication it can be minimized. This may be a more appropriate plan if there is not enough staff available at the damaged library. Processing the incoming materials will also be a logistical aspect to consider.

GETTING HELP TO PREPARE YOUR INSTITUTION

Given the amount of work required to prepare and institution, where does one begin–especially in the many smaller institutions in which low staffing requires that administrators serve in many different roles? The answers generally fall into three primary categories–grants, professional advice, and self-education.

Grants

Grants are available to assist institutions evaluate preservation needs and develop disaster plans from a variety of sources. The nation's largest program focused exclusively on providing smaller institutions with preservation assistance is run by the National Endowment for the Humanities (NEH). Known as Preservation Assistance Grants, the NEH's program provides institutions with funding of up to $5,000 for preservation needs assessments, consulting, storage improvements, and treatments. Many institutions use these grants to offset the costs of hiring consultants to develop disaster plans, equip themselves with disaster supplies, or both. The application process is designed to be relatively simple, and the program is intended specifically to reach smaller institutions.

In many states, the state libraries provide funding through their Library Services and Technology Act (LSTA) grant program or through other state-funded programs that can be used for preservation and conservation services. Likewise, other national, state, local, and private organizations provide funding that can be used to assist in the preparation of a disaster plan, in mitigation activities, or in the development of a local institution's capacity to protect and preserve its collections.

Professional Advice

Staff at universities, particularly at land-grant state institutions, take their public service mission very seriously. Often preservation and conservation professionals affiliated with these libraries provide free advice to the public.

Riding on the wave of interest in preservation and conservation in the 1980s and early 1990s, preservation service centers developed that provide advice about preservation. Some, like the Northeast Document Conservation Center, are independent entities. Other resources include preservation services at networks like Solinet and Amigos. In many cases, these regional service centers include not only online resources and field service officers that provide advice on preservation and conservation needs, but also 24-hour disaster response hotlines, training programs, and grant writing assistance.

Educating Yourself

Perhaps the most important activity that librarians can pursue to protect their collections is to educate themselves. As the preservation profession developed, most communication about best practices in preservation and conservation centered on newsletters, published papers, conferences, and workshops. Today, the public at large has access to a plethora of information. The Internet's growth has put preservation information at the tips of our fingers, and email distribution lists of preservation information provide information to even the most passive users. Preservation service centers maintain extensive online collections of preservation information. Likewise, national organizations like the American Institute for the Conservation of Art and Historic Artifacts and national institutions such as the Library of Congress and the National Archive and Records Administration provide extensive online collections of materials related to preserving collections.

In addition to providing resources that can assist institutions in preparing themselves for disasters, Federal and state agencies also provide resources to help institutions recover from a disaster. The Federal Emergency Management Agency and various statewide emergency management agencies offer invaluable assistance to qualifying institutions located within declared disaster zones. In some cases, these agencies provide funding to support mitigation activities as well. However, knowing about these activities requires educating oneself prior to a disaster.

TAKING ACTION–DEVELOP A PLAN

In addition to online resources, continuing education programs provide an outlet for librarians to receive professional training in preservation. Working through their field service officers or consultants, regional service centers frequently offer face-to-face and online training programs in specific preservation topics.

The Internet has also made different types of resources available. One of the most recent is dPlan, <http://www.dplan.org/> (NEDCC, 2006). Developed through funding from the Institute for Museum and Library Services, dPlan is an online tool developed expressly for the production of disaster plans. Although using the program does require registration, the tool provides a fool-proof method for developing a disaster plan that will serve a local institution.

In addition to planning locally, smaller institution may benefit by collaborating with other nearby libraries or institutions. Such arrangements may be formal or informal. They may require drawing up written contracts of memoranda of understanding, but the benefits of having assistance in a time of need is not to be understated. In small towns or multiple occupancy facilities, building such cooperative relationships before a disaster strikes can make the difference in salvaging the most important documents within a collection.

CONCLUSION

Finding the right people (locally, regionally, and nationally) to support your library, creating and updating a disaster plan, and learning about how to conduct a disaster operation can make all the difference in the world in salvaging the resources within your care. Taking the time now to discuss with the fire department, state library, or regional disaster vendor could save time and money in the event of a disaster. Creating a disaster plan can be a relatively short-term project, and will provide guidance if it's ever needed. There is a large amount of information available for more research and the key is to not "plan for the worst" but "plan for everything." There are articles about specific types of disasters. There are articles about specific types of libraries and archives. Look for appropriate or easily adaptable information. Reading a couple of articles about specific disasters provide a general idea of a disaster situation, and the chronology of what happens to the staff, building, collection, and community. Books that focus on disaster preparedness will provide a lot of the guidance needed to prepare. However, it is important to remember that no matter how well prepared one's institution is prior to a disaster, it is only when the disaster strikes that individuals will rise or fall.

The differences between a disaster at a large library and at a small library are both positive and negative. A large library may have optimal amounts of staff and resources to respond to a disaster, but the damage may be far greater. A small library may not have space or money to have a large disaster kit, but according to Ashbaugh "there may be a more direct level of citizen involvement" (2003, p. 7), which can provide more support and resources than professional response.

REFERENCES

Ashbaugh, Ellen. (2003). The Effect of Disasters Upon the Rural Library. Rural Libraries 23(1), pp. 7-20.

Bolger, Laurie. (2003). Scared or prepared? Disaster planning makes the difference. Information Outlook, 7(7), pp. 26-30.

Clareson, Tom et al. (2005). Illinois statewide preservation survey project: 2005 preservation planning survey report. Dublin, OH. Retrieved December 8, 2006 from <http://www.library.uiuc.edu/prescons/grants/Illinois%20Statewide%20Preservation%20Survey%20Report.pdf>.

Fortson, Judith. (1992). Disaster Planning and Recovery: A How-To-Do-It Manual for Librarians and Archivists. New York: Neal-Schuman Publishers.

Heritage Preservation. (2005). A Public Trust at Risk: The Heritage Health Index Report on the State of America's Collections. Washington, DC. Retrieved December 8, 2006 from <http://www.heritagepreservation.org/hhi/>.

IFLA/UNESCO. (1999). Survey on digitization and preservation. Wetherby, United Kingdom: IFLA Offices for UAP and International Lending.

Kane, Kim. (2001). You Too Can Do a Disaster Plan. Library Mosaics 12, pp. 12-13.

Matthews, Graham and Feather, John. (2003). Disaster management for libraries and archives. Burlington, VT: Ashgate.

Northeast Document Conservation Center. (2006). Retrieved December 8, 2006, from <http://www.nedcc.org>.

Northeast Document Conservation Center. (2006). dPlan: The online disaster planning tool. Andover, MA. Retrieved December 8, 2006 from <http://www.dplan.org/>.

Wellheiser, Johanna and Scott, Jude. (2002). An Ounce of Prevention: Integrated disaster planning for archives, libraries, and record centres. Lanham, MD: Scarecrow Press.

doi:10.1300/J118v25n03_05

Fire at University of New Mexico Library

Terry Gugliotta

SUMMARY. A fire in the basement of the University of New Mexico's Zimmerman Library destroyed 30,000 volumes of periodicals as well as furnishings, shelving, and fixtures in the path of the flames. An earlier flood in the underground science and engineering library had prompted the creation of a disaster plan and the formation of a recovery team. As a result, the library was prepared to handle a disaster of the magnitude of the fire. Damage estimates are put at $10-15 million not including the cost of staff time lost. Although the state fire marshal has acknowledged that the fire was deliberately started with an accelerant, he has not yet officially ruled the fire as arson. doi:10.1300/J118v25n03_06 *[Article copies available for a fee from The Haworth Document Delivery Service: 1-800-HAWORTH. E-mail address: <docdelivery@haworthpress.com> Website: <http://www.HaworthPress.com> © 2006 by The Haworth Press, Inc. All rights reserved.]*

KEYWORDS. Fire suppression sprinklers, Disaster Recovery Assistance Team, Emergency Preparedness and Recovery Plan, disaster recovery, smoke damage, phone tree, interlibrary loan, electronic resources

---

Terry Gugliotta, BA, is Archivist, University of New Mexico Library, Albuquerque, NM 87131 (E-mail: tgug@unm.edu).

[Haworth co-indexing entry note]: "Fire at University of New Mexico Library." Gugliotta, Terry. Co-published simultaneously in *Public Library Quarterly* (The Haworth Information Press, an imprint of The Haworth Press, Inc.) Vol. 25, No. 3/4, 2006, pp. 61-69; and: *Dealing with Natural Disasters in Libraries* (ed: William Miller and Rita M. Pellen) The Haworth Information Press, an imprint of The Haworth Press, Inc., 2006, pp. 61-69. Single or multiple copies of this article are available for a fee from The Haworth Document Delivery Service [1-800-HAWORTH, 9:00 a.m. - 5:00 p.m. (EST). E-mail address: docdelivery@haworthpress.com].

Available online at http://plq.haworthpress.com

doi:10.1300/J118v25n03_06

DISASTER STRIKES

At 10:40 p.m., on Sunday, April 30, 2006, fire swept through the basement of the University of New Mexico's historic Zimmerman Library. The flames destroyed 30,000 bound periodical volumes, melting light fixtures, and steel shelving in their path.

The initial call was a two-alarm fire and Albuquerque's Engine Company #3, located on the eastern edge of the campus, arrived within minutes. Thick, black smoke was already billowing from the building. The fire, which reached temperatures of 1,400 degrees Fahrenheit, spread quickly, but everyone had been safely evacuated according to library safety procedures.

UNM police were also at the scene within minutes. They met the library's night supervisor, Russ Cole, who confirmed the building had been evacuated. Police rapidly secured the building, blocking every entrance and likely saved the lives of students who attempted to re-enter the building to retrieve their laptops and other belongings.

Meanwhile, Cole explained to firefighters where the fire originated. UNM Fire Marshal Vince Leonard said although there was substantial loss of property, victory over the fire was secured faster because Cole showed firefighters several additional routes into the basement, which allowed them to gain access to the fire.

UNM's team of first responders, physical plant, and environmental safety employees, were also among the first to arrive at the scene with "as built" drawings in case firefighters needed to locate electrical shutoff switches.

Firefighters broke basement skylights to let smoke escape, then entered the building, dragging hoses to the basement. But the fire was already too big and too hot. The firemen retreated. Two additional engine companies were ordered to the scene. The library was built before fire suppression sprinklers were mandatory and Zimmerman did not have them.

UNM Police Chief Kathy Guimond was notified by dispatchers at 10:45 p.m. She tracked down Library Dean Camila Alire in Washington, D.C., then called Associate Dean Fran Wilkinson to alert her. Wilkinson grabbed her Emergency Preparedness and Recovery Plan, which all members of the library's Disaster Recovery Assistance Team (DRAT) keep at home, and headed for campus.

The library had prepared for the unthinkable by compiling an emergency plan in the mid-1990s in response to flooding in the underground science and engineering branch library. The purpose of the first emergency plan was simply to determine salvage priorities of each department and to locate emergency shut-off switches and supply closets. After the next flood, it became apparent that maintaining the emergency contact list was critical. Key library personnel identified included facilities staff, human resources staff, branch directors, and department heads. The Emergency Preparedness and Recovery Plan is updated frequently.

Wilkinson arrived on campus and met Guimond and Public Affairs Director Susan McKinsey. On her way to campus, Wilkinson called library Facilities Manager Ed Padilla, a member of DRAT, who also rushed to campus. Chief Guimond called the provost, who was the ranking UNM official at the time, while McKinsey briefed reporters.

Albuquerque firefighters had initial jurisdiction over the scene. They set up a command station for people to check in and out and to release information to University officials. When firefighters re-entered the basement with additional crews the smoke was still so thick they had to crawl on their bellies, dragging hoses behind them, to find the flames. As they approached the flames they smelled an accelerant. UNM Fire Marshal Leonard called the state fire marshal who dispatched seven investigators to the scene.

Because officials suspected arson, there were strict procedures related to command and control of the building. The fire was contained in two hours, but the building remained under the control of the firefighters until hot spots were cooled and smoke was cleared from the building.

The state fire marshal took control of the building at 2 a.m. It was evident that the library would not reopen soon.

The investigators sifted through the basement, quickly locating the origin of the fire confirming that an accelerant was used to start it. At 4 a.m. the state fire marshal took Wilkinson, Padilla, Guimond, and Provost Brian Foster on a tour of the building. Water was dripping everywhere in the basement. Ceiling tiles were missing. The smell of smoke was strong. They were only able to view the burned area from a distance, but they saw thousands of burned and partially burned volumes of periodicals. Shelving leaned 45 degrees with wet books that had been lapped by flames. Everything was wet. The investigation continued for the next week. No disaster recovery teams were allowed into the building, leaving volumes to sit in water. Many were destroyed.

Smoke was carried to nearly every corner of the building through the ventilation system. In the reference area on the first floor above the fire, the walls were streaked with soot. The second and third floors were also covered with a light dusting of soot. Smoke hung in the air of the west side of the building, which included Special Collections. The nine-story book tower, which holds much of the rare book collection, had some smoke damage, while the three-story underground vault that holds the archives had the least damage.

MAINTAINING CUSTOMER SERVICE

Even before the fire was extinguished, Wilkinson's primary concern shifted to developing a plan for the library to maintain service to students. It was the week before finals and students would need library materials to finish research papers and space to study for exams.

As investigators searched for evidence, Wilkinson stood at the edge of the roped off fire scene and began to plan. Using her personal cell phone, she called the director of the Student Union Building (SUB) and arranged to set up reference services there and to have the SUB extend its operating hours so students would have a place to study. She then contacted branch directors who also extended their library's hours. The directors were also told to activate the phone tree in the Emergency Preparedness and Recovery Plan. They called department heads from Zimmerman Library who then called their employees. Wilkinson sent an E-mail to all library employees apprising them of the situation. Police Chief Guimond sent a campus-wide message to all faculty, staff, and students via a special E-mail emergency alert system the University had set up in 2005. The library's web page and an emergency phone line also provided daily updates of the situation.

Library payroll was due Monday, May 1, and all of the files and computers containing critical information were on the second floor of Zimmerman. The fire marshal allowed Facilities Manager Ed Padilla to retrieve the materials needed to get the payroll finished.

Students needed materials professors had placed on reserve. They would also need journals, books, and reference materials. Mailroom and circulation department employees moved reserve materials and interlibrary loan operations to branch libraries on Tuesday and Wednesday. Paging books from the main collection, located on the second and third floors, began later that first week. Books were delivered to the science and engineering library for patrons to check out. Wilkinson and others notified the Association of Research Libraries, the New Mexico Library Association, the New Mexico Academic Library Consortium, and interlibrary loan partners to alert them to the situation and ask that UNM's request to borrow materials be made a priority. Students and faculty received journal requests electronically within 24 hours of their requests. Throughout the University, professors worked with their students to make arrangements for late papers and to modify assignments affected by the fire.

The library set up information tables outside of the entrances to Zimmerman on the first day to let customers know there had been a fire and the building was closed. Some displaced Zimmerman employees staffed the information tables and the reference tables at the SUB.

DISASTER RECOVERY

Since the library had experience with flooding in its science and engineering branch and had worked with Rockefeller Cleaning Services, the company began the initial basement cleaning May 11, the day after the University was granted temporary occupancy by the state fire marshal. Maloy Construction Company was hired to stabilize shelving and inspect the building to ensure its integrity before Rockefeller began its work.

On May 11, UNM also hired BMS CAT Ltd., a company specializing in disaster recovery, for most of the removal of the debris, the pack-out of salvageable items, the decontamination of the HVAC system, and the cleaning of documents, books, and furnishings.

BMS CAT sent 35 employees from their regional office in Dallas, Texas to manage the job, which included hiring 100 additional temporary local employees. The basement houses journals, newspapers, government documents, microfilm, and microfiche collections. In 13 days, workers removed the majority of the scorched remains of books, shelving, and furniture, filling approximately 60 construction dumpsters. They also packed out 15,000 boxes of books and 125 microfilm cabinets. The basement resembled an ant colony with workers going in every direction focused on their jobs. Crews worked two shifts seven days a week.

Dan Barkley, director of Government Information, which was located in the basement, volunteered to pore through damaged volumes to make lists of material that could be salvaged. Barkley and Librarian Anne Schultz inspected 70,000 volumes in just a few days. Because the fire destroyed the main electrical closet, they used mining helmets with lights to see what they were doing. Barkley labeled shelving and drew a map to record the location of the books packed out. He and Schultz also inspected all of the microfilm and fiche, which had survived the inferno but was contaminated by smoke and soot, before it was packed out.

Three weeks after the fire, BMS CAT allowed library staff to retrieve personal items from their offices. They gave staff protective gear and brought them into the basement in groups of fifteen. BMS CAT also provided information about how to clean and deodorize smoke damaged belongings. Few, if any personal items were lost. Some plants died, but Padilla had been on site assisting BMS CAT and had managed to water the plants he saw.

The DRAT team expanded to include volunteers, UNM safety, building and construction personnel, and members of BMS CAT. They met every day at 7 a.m. and 2 p.m. to review the day's work and plan for what still needed to be done. People were given the freedom and flexibility to make on-the-spot decisions to get things done.

Since smoke had reached nearly every corner of the building, workers cleaned it from top to bottom. Every ceiling tile was replaced. Every wall, light switch, bathroom fixture, book, shelf, chair, table, floor, and window was cleaned throughout the building. Fans and air scrubbers helped reduce the smell of smoke and the entire HVAC system was cleaned, including scrubbing the insides of the duct work. After the periodicals, newspapers, government documents, microfilm, microfiche, and furnishings were packed out, the building was treated overnight with ozone to help remove the smell of smoke. BMS CAT finished its work at the end of July. In all 17 semi-trucks of periodicals, microfilm and documents were sent to Texas to be cleaned and stored until the basement is rebuilt.

After the pack-out and cleaning, Maloy Construction Company returned and demolished the interior walls in the basement and on the first floor above the fire and removed carpeting and fixtures. They also chipped the inner surface of the basement level concrete walls to remove smoke odors. The walls were then sealed with primer.

Throughout the clean-up, University safety officers monitored air quality and safe work practices. Just as no one was injured in the fire, no one was injured in the clean-up.

MODEL

If good can come from the tragedy, it has come in the form of teaching and awareness. University Fire Marshal Leonard and Safety Director Bob Dunnington have collected hundreds of photos of the scene as well as examples of burned items, including smoke detectors, to put together a new fire safety presentation to use on campus and to share with the Albuquerque Fire Department.

In the first week following the fire, rookie firefighters were brought to the scene for review sessions. Zimmerman did not have fire suppression sprinklers and, although an accelerant was used to start the fire, it did not spread as fast as it could have because the books were packed so tightly on the shelves that fire could not get inside them quickly. In addition, the multiple access points pointed out by Russ Cole allowed firefighters to attack the fire from three sides.

Because Zimmerman had an up-to-date Emergency Preparedness and Recovery Plan, from the first moment employees had a road map toward recovery. Ironically, multiple small construction projects in the library during the previous 15 years were instrumental in assuring the success of the emergency plan. These projects frequently set off fire alarms that forced the building to be evacuated, allowing library staff to practice fire drills and to refine plans as necessary. Since planned fire drills are costly, the library quickly began to take advantage of these unplanned drills and designed a detailed evacuation plan. All of those accidental fire drills saved lives in the real emergency.

Fire safety training for faculty, staff, and students across the University is cost-prohibitive. UNM Fire Marshal Leonard wishes he had the staff and funds to assist each department in developing an emergency plan like the library's. He provides safety training and fire drills to departments by request. The remainder of his time is spent reviewing fire evacuation plans and inspecting more than 200 buildings that make up the Albuquerque campus.

AFTERMATH

In the early hours of the fire, Wilkinson thought the library might be closed for a few days, maybe for a week. In the end, it was closed for more than two months.

In all, 104 employees were displaced because of the fire in Zimmerman. It destroyed the basement as well as the reference area on the first floor, ruined 30,000 volumes of material, damaged computers, printers, microfilm readers, and copy machines, and destroyed the entire building's fire alarm system.

A handful of employees returned to their offices in the first week to assist remediation crews. The rest were assigned to branch libraries or worked from home. Some employees were able to continue their projects, but most were assigned new projects in the branch libraries.

The alarm system throughout the building had been destroyed because many parts came together in an electrical closet in the basement corner where the fire erupted. A little more than a week after the fire, when the state fire marshal granted the University a temporary occupancy permit, the library put forth a "fire watch" plan and trained employees to do the job. This allowed the library administration and the Budget and Cost Management Department to move back into the building. Padilla divided the library into zones and assigned staff to walk through those zones to make sure there were no fires in the building. Padilla issued air horns, hard hats, maps, and keys to staff. The fire remains under investigation and has not been officially ruled as arson. All leads are being investigated, but Police Chief Guimond says it will be a hard crime to solve since there are no clues from the suspect such as threatening letters or graffiti. There were also no witnesses or security cameras in the area.

Sprinklers have been added to the basement and the first floor area now under reconstruction. The library will also increase its number of security guards and staff working during late night hours. Security cameras and compact shelving will be added as additional funding is approved. The library has taken this opportunity to reconfigure the first floor reference area and the basement to accommodate the way students study today. More group study rooms and group computer stations are being designed so students can work more collaboratively.

Valves for the sprinkler system will be located behind locked doors and the valves will also have tamper switches to prevent an arsonist from turning them off. The HVAC dampers will be equipped with smoke detectors to prevent the spread of smoke through the building.

The first floor reference area reopened in January 2007. The basement may reopen in December 2007. Until then, faculty and students will have to rely on interlibrary loan and electronic resources that Zimmerman is now purchasing to replace lost journals. Linda Lewis, collection development coordinator for Zimmerman and its branch libraries, is heading efforts to replace journals. She says that offers have poured in from around the country to replace anthropology and southwest history titles, but many of the Latin American materials that were lost are not replaceable. One hundred years of collecting in these subject areas burned in less than an hour. Lewis says that although just 20% of the collection was lost, it is the most difficult part to replace. The fire occurred just before all journal invoices were due for payment. Lewis contacted vendors and explained that the library not only was unable to locate the invoices, which had been packed out, but also did not have access to the equipment that employees needed to generate the payments. Vendors were understanding and waited for payments. Several even gave the University free access to databases that included some journals lost in the fire. Others have offered back issues of titles that were lost.

Initial estimates of the total damage from the fire were $5 million. Today, estimates are $10-15 million. The fire has affected every one differently. Illene Renfro, serials cataloger for 30 years, thought reports of the fire were a joke at first. She was in disbelief when she learned it was true and that so much of her life's work at the library had gone up in smoke. Faculty and students who use the anthropology and Latin American materials are upset at the loss since they depended heavily on many of the journals that were destroyed.

Lewis continues to struggle with the question of why a person would ruthlessly destroy this intellectual knowledge. And though the fire has truly been a tragedy, part of the silver lining is the increased support from the community. So many grew up on the campus with parents teaching at the University or earned their degrees at UNM. The outpouring of support from those who love the library has been a tremendous help in the process of healing.

doi:10.1300/J118v25n03_06

The Great University of Georgia Libraries Fire of '03: Lessons Learned and Questions Raised

Ryan Perry

SUMMARY. This essay describes a fire at a major state institution that caused primarily fire-related damage, with very little attendant water damage. The essay covers treatment of materials damaged and disaster plans as they relate to an incident of this magnitude, and suggests ways to assist other institutions in dealing with major disasters requiring contracted assistance. doi:10.1300/J118v25n03_07 *[Article copies available for a fee from The Haworth Document Delivery Service: 1-800-HAWORTH. E-mail address: <docdelivery@haworthpress.com> Website: <http://www.HaworthPress.com> © 2006 by The Haworth Press, Inc. All rights reserved.]*

KEYWORDS. Disaster plans, disaster recovery, fire damage, preservation, conservation, disasters

---

Ryan Perry, BA, is Preservation Archival Associate, University of Georgia Libraries, Athens, GA 30602 (E-mail: rperry@uga.edu).

[Haworth co-indexing entry note]: "The Great University of Georgia Libraries Fire of '03: Lessons Learned and Questions Raised." Perry, Ryan. Co-published simultaneously in *Public Library Quarterly* (The Haworth Information Press, an imprint of The Haworth Press, Inc.) Vol. 25, No. 3/4, 2006, pp. 71-88; and: *Dealing with Natural Disasters in Libraries* (ed: William Miller and Rita M. Pellen) The Haworth Information Press, an imprint of The Haworth Press, Inc., 2006, pp. 71-88. Single or multiple copies of this article are available for a fee from The Haworth Document Delivery Service [1-800-HAWORTH, 9:00 a.m. - 5:00 p.m. (EST). E-mail address: docdelivery@haworthpress.com].

Available online at http://plq.haworthpress.com

doi:10.1300/J118v25n03_07

INTRODUCTION

At approximately 5:40 p.m. on July 23, 2003 someone set a fire on the second floor of the Ilah Dunlap Little Memorial Library at the University of Georgia.[1] There were two, possibly three, points of origin: a stack of approximately two hundred boxes along the wall to a storage room; a ceiling tile along the south outside wall by the first range of books; and an end of a shelf within the book ranges themselves (Illustration 1). The fire spread to the ceiling tiles from the boxes, spread in a pattern emanating out from a point just above the microfiche cabinets, and rained down on the assorted materials below. There were, at the time, no sprinklers on the second floor, as is the case in many buildings dating from the 1970s or earlier.

In the meantime, due to the intense heat of the fire, massive quantities of smoke were produced and sucked into the building air-conditioning system, circulating the smoke first up to the seventh floor and then down from there. The fire alarms began going off, first on the seventh floor, and then the sixth, and then the fifth, and so on, causing the firemen to rush to the seventh floor first upon entering the building. The fire, because it spread primarily between the drop-ceiling tiles and the

ILLUSTRATION 1. Floor Plan of the Second Floor Annex

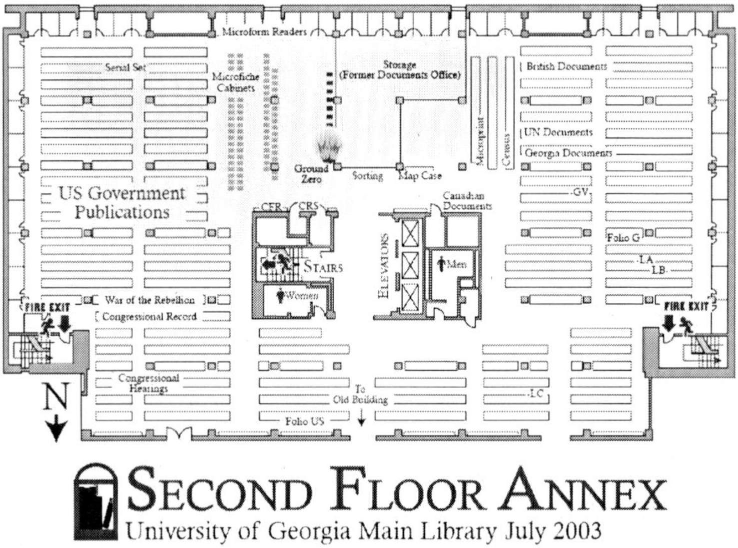

Reprinted with permission.

cement ceiling, largely consumed the oxygen available to it, and therefore spent itself, before the firemen arrived to extinguish it.

The fire, which burned for about twenty minutes, ended up costing several million dollars and countless hours of manpower thus far, with recovery efforts still ongoing. It produced, moreover, damage of a sort not much discussed in any of the preservation literature dealing with disasters, as well as a situation that could not possibly have been dealt with by the regular library staff, complicating the advice given in the preservation literature as to the implementation of a disaster plan.

Various factors involved in the fire damage and recovery efforts led to an unusual set of problems for the materials affected by the fire, problems caused solely by fire damage. By and large, the relevant preservation manuals advise preservationists on what to do in cases where materials have been scorched by a fire and then doused by those attempting to extinguish the fire; hardly ever do these manuals consider damage caused solely by fire.[2] By discussing what happened in this one particular instance, this essay is an attempt to open up a topic under-represented in the preservation field, though one that could potentially affect a very large number of organizations.

For the most part, disaster plans tend to focus on what an organization can do in the case of a disaster of a size that the organization itself can handle.[3] The University of Georgia Libraries fire was obviously a disaster of a larger order and, as such, the Libraries had to seek the assistance of a professional disaster recovery company. Using the experience of this specific disaster as a test-case, this essay raises questions regarding the nature of disaster plans, their effectiveness, and problems involved in their implementation.

WHAT WAS BURNT?

The most important factor in understanding the types of fire damage is location. The Ilah Dunlap Little Memorial Library is the main facility of the University of Georgia Libraries and it houses 2.6 million volumes, including all of the humanities and social science Library of Congress classifications as well as some Government Documents; the second floor at the time of the fire also housed some Government Documents, and the A, C, LA-LC, GT, GV, and Z Library of Congress classifications, as well as a collection of University of Georgia theses and dissertations, a total of approximately 100,000 volumes and 450,000 Government Documents. The first four floors of the building as well as the basement and sub-basement are divided into the "old building," constructed in 1952, and an annex, constructed in 1974, with the two parts separated by a fire door on each floor. The fire was contained on the second floor, on the annex side, although the smoke damage spread throughout the building.

The fire began with a group of almost two hundred boxes that contained UN documents brought back from their temporary housing at the University of Georgia Libraries remote storage facility (the Repository) during a heating and air-conditioning renovation project. The fire quickly swept through the boxes and spread to the ceiling tiles, radiating from its main origin point. From the ceiling tiles, the fire rained down on the material below, including several ranges of print Government Documents (including the U.S. Congressional Serials Set) on the east side of the floor, all of the Government Documents microfiche cabinets (which were located between the boxes and the stacks), a cabinet holding Government Documents on CD, the map case for Government Documents, Microprint boxes on the far side of the Storage room, and the computer hardware located in the storage area. The fire, because it was raining down on the materials from the ceiling tiles, created a pattern whereby the material nearer to the ceiling received the worst damage and the material closer to the floor received less (Illustration 2).

This pattern and the wide array of material it affected occasioned several different kinds damage:

1. Burnt and Wet. This category included the smallest number of damaged items. The items in this category had been charred to a great extent by the fire and then had been doused with water and allowed to sit.[4] One should note as well that, by and large, the fire had consumed itself by the time the firemen arrived, so the amount of water they had to use to put out the fire was minimal. In addition, the Athens-Clarke County Fire Department trains its employees on the proper handling of archival material; hence, the firemen were extremely careful about the amount of water they were using. The care that the Fire Department took with water management meant that primarily only the material in the boxes used to start the fire and the material on the sorting shelves nearby the main point of origin were affected.
2. Completely Burnt. There were a few items that had been significantly charred by the fire, which reduced a large portion of their mass to ashes. These materials were mostly microfiche in baskets on top of the microfiche cabinets awaiting filing, paperbound items

ILLUSTRATION 2. Ceiling-to-Floor Burn Pattern

Photo courtesy of UGA Libraries, Katie Gentilello. 2003. Reprinted with permission.

on the top portion of shelves (which were hardest hit by the fire), including paperbound items on the lower shelves in which a significant amount of content was burnt, bound items on the very top shelves which were brittle, and items in the boxes that were used to start the fire (Illustration 3).

3. Partially Burnt. Of the materials that were directly in contact with the fire, this category represents the largest number. Items in this category had only their edges or casings charred by the fire. It includes paperbound items found on the lower shelves in which most of the content remained intact, bound items on the top shelves and on the lower shelves, and microprint materials.
4. Heat Damaged. The materials in this category suffered the greatest variety of damage. Some items suffered intense heat within their containers, such as the CD-ROMs in the CD case, the microfiche, some of the microprint, three-ring notebooks, and spiral bound material (Illustration 4); other items were affected indirectly by the heat, as in cases where the plastic light fixtures or item markers

ILLUSTRATION 3. The Boxes Used to Start the Fire

Photo courtesy of UGA Libraries, Katie Gentilello. 2003. Reprinted with permission.

melted and fell onto the material underneath (Illustration 5). Also included in this group is all of the computer equipment both out on the floor and in the storage office, including quite a bit of computer hardware. The damage in this group was unique to each item and, as such, will not be treated at length in this essay, although general trends will be discussed.

5. Soot Damage. This category represents the largest number of items by far, and can be divided into two subcategories: the items on the annex of the second floor, which suffered the heaviest soot damage, and the rest of the items in the building, which suffered lighter soot damage. All of the items in the Main Library were covered with soot. After the smoke was sucked into the air conditioning system and distributed to each of the floors after the air conditioning had filtered out some of the heavier soot particles, the thickest layer of soot was on the seventh floor and then the soot damage decreased in severity by each floor. The materials on the second floor, on the Annex side with the fire, were an exception to

ILLUSTRATION 4. Varieties of Heat Damage

Photo courtesy of UGA Libraries, Katie Gentilello. 2003. Reprinted with permission.

this scheme of decreasing severity. These items were covered with a very thick layer of soot, which had become a kind of sludge, due to the large number of plastic items that were consumed by the fire.

Obviously, there was a large variety of damage done to items and an enormous number of items affected by the fire, complicating the cleaning process and necessitating a disaster recovery service. The first step in the recovery effort was to identify the kinds of damage listed above, allowing the Preservation Department to decide what materials could be handled by a disaster recovery service, and what materials about which the librarians would have to make decisions. The next section of the essay will detail how the material was sorted, who did the sorting, and what was done to each of the types of damage.

ILLUSTRATION 5. Varieties of Heat Damage

Photo courtesy of UGA Libraries, Katie Gentilello. 2003. Reprinted with permission.

WHAT WAS DONE?

The first two steps of the fire recovery effort were to assemble a team of library staff that would be able to make competent decisions regarding the collection and handle materials with which the disaster recovery service would not have the appropriate training and to find a disaster recovery service that could handle the appropriate problems. In this instance, the fire recovery team consisted of the preservation department, which included a preservation librarian, a preservation assistant, and two trained student workers, as well as the government documents staff, consisting of the federal government documents librarian, the U. S. Regional Depository librarian, the state documents librarian, the Census collection librarian, and three government documents assistants, along with a large number of library staff volunteers.[5]

There was no need to involve the bibliographer who oversees the Library of Congress classifications on that floor because those items were not directly affected by the fire, and all items suffering soot damage only were treated in the same manner. Preservation workers were needed so that they could make decisions regarding what could be saved and what was unsalvageable. The government documents workers were needed so that they could quickly and accurately identify the portions of the collection that had sustained the greatest amounts of damage. The Libraries also needed a contract with a disaster recovery service, so that the process of working out the logistics of such a large project could begin. In this case, a disaster company already under contract for the state of Georgia was on campus to assist with a flood at the College of Veterinary Medicine. The decision of what disaster company to use was thus not made by the library, but by the University's administration and the representatives of the state government.

Once the fire recovery team was assembled, and the disaster recovery service had been contracted, the task of sorting the damaged material could begin in earnest. It must be stated, however, that there was an initial delay to the recovery efforts. Since the fire was ruled as arson, the second floor was declared a crime scene and so no recovery personnel were allowed to begin work until the law enforcement agencies had inspected the site. The police department understood the conservation concerns and so worked very quickly, only causing about a day's delay. However, since the delay was in the middle of July in Georgia, it created an ample opportunity for mold to take root in some places.

Once the recovery team was able to assess the situation, it divided the work between the library recovery team and the disaster recovery service. Since the preservation team decided that all of the soot-covered material could be treated in the same way, and because that treatment did not require any specialized training, the disaster recovery service was instructed to handle what was by far the bulk of the work. For the items on the second floor that had been directly affected by the fire, the paper media and the other media were treated differently.

As luck would have it, there was a cluster of three ranges of shelves that were empty and awaiting shifting, immediately after the U.S. Congressional Serial Set, in the area that had sustained the majority of the damage, allowing the recovery team to create a triage area. The items that were either completely burnt, partially burnt, or heat damaged were moved into this area, so that the library staff could take a closer and longer look at each item. The materials moved into this area came primarily from the top shelves in the area affected by the fire (represented by shaded area in Illustration 1), because the materials on the lower shelves were merely soot damaged. The other media affected by the fire were packed and moved to an alternate remote storage facility designated by the University to house the material from the second floor annex, while the second floor was being rebuilt. As the rest of the building was being cleaned, and the material on the second floor was being relocated, the staff was able to make decisions regarding how to best handle the material that had been affected the worst.

CONSERVATION STRATEGIES

1. Burnt and Wet. The items on the sorting shelves and the items in the boxes were completely lost. What was left of the items after they had been burned was then soaked with water and then allowed to sit in an environment that was not climate controlled while the police investigated the arson. So, not only had the vast majority of the mass of the items been consumed by fire, but also what was left had been covered by mold. These items were quickly removed from the library. The items in the boxes were all considered destroyed and the items on the sorting shelves were sorted through, in order to see what was salvageable, though nothing was, and to identify specific documents whenever possible. Perhaps the most important items destroyed in this group were two volumes of the *American State Papers*.
2. Completely Burnt. Once the materials that had been directly affected by the fire were moved into the triage area, the preservation staff had the task of deciding what items could be saved and what items were too burnt to be viable for future use. This task involved looking at each item individually, since some smaller pamphlets shelved between larger pamphlets escaped being burnt at the expense of the larger pamphlets. Likewise, some material housed in manila envelopes escaped being burnt, while the envelopes themselves were partially consumed by the fire. The preservation team then attempted to determine a range of call numbers in which all the materials in that range were destroyed. The practice of determining call number ranges was one of the mistakes that, in hindsight, could have been dealt with more efficiently. However, the items that were completely burnt presented a unique challenge in

the task of determining call numbers: how does one go about determining the classification for an item that is illegible, or completely reduced to ash? The items that were completely burnt were declared lost and reported to the insurance company.

3. Partially Burnt. Apart from the microprint, the items in this category can be divided into two categories: items too small to bind and items large enough to bind. The microprint was taken out of its case, sometimes using a hammer, due to the fact that the fire had caused the plastic casing to fuse, and each piece of microprint was examined to determine whether or not any of its content was damaged. Those that were not, were treated like the soot covered items and placed in replacement cases. The print documents that had been partially burnt, but had sustained no significant loss of content, were left on the shelves in the triage area; these items would be treated like the soot covered items and then placed into acid free envelopes. The print documents that were large enough to bind, and the already bound items, were sent to a commercial bindery that handles routine binding for the libraries, so that the items could be rebound. The recovery team met with the owner of the University of Georgia's contracted bindery and arranged to send the bindery the partially burnt items, with the minimal amount of information necessary to mark the items properly. The government document workers had the task of reading the call numbers, or using the accession numbers to generate the call number, and then filling out a binding slip, so that the bindery could process the documents properly. In addition, the workers had to determine whether or not the item was brittle and so could not be rebound, or contained foldouts, in which case the item could not be trimmed. The brittle items were treated like the soot-covered material and then placed into protective enclosures (or phase boxes).
4. Heat Damaged. The items in this category were unique in the kind of damage they sustained. In general, the recovery team attempted to pry the molten plastic off of the print media, without letting the plastic take too much of the paper with it. The item could then be either rebound, or treated like a soot covered item. The microprint items were treated in the same manner as the paper items, with the exception that the microprint boxes had to be replaced by acquiring used microprint boxes from Emory University. The microfiche and the CD-ROMs were moved to the alternative remote storage facility, where they remained until they were cleaned of

soot and then tested to see if they were still usable. Many of the microfiche that had been on the top shelves of the microfiche cabinets and the CD-ROMs and microfiche that had been near the front end of the drawers closest to the fire were too damaged by the heat to be recovered. All of the computer equipment was damaged beyond repair.

5. Soot Damaged. This category of material, the largest, was treated in two different ways, depending on its location within the building. All of the materials in the building, excluding materials on the second floor annex, were cleaned in the library. The disaster recovery company handled each item individually, gripping it tightly so that the soot did not fall between the pages, and wiping the items with a dry chemical sponge.[6] The items on the second floor annex, which had to be moved so that reconstruction could begin on that floor, were packed tightly spine down to insure that soot would not fall between the pages, and then sent to the alternate remote storage facility, where they were treated like the soot-covered items in the rest of the building. The disaster recovery company was responsible for packing and moving the items.

The number of items that were declared destroyed had to be revisited several times due to the inaccuracy of declaring entire call number ranges completely burnt. It must be stressed, however, that this is only a fraction of the items that might have been lost, had not the recovery team decided to rebind the vast majority of burnt items. And of course, since every item in the Main Library was affected by the fire, the University of Georgia Libraries could not throw out every item that was affected. The recovery team felt that, rather than declare all affected, or directly affected, items a loss, especially since several items could not be easily replaced, it was necessary to try and simply repair the items that could still be useable. The second floor was rebuilt and a sprinkler system installed in the second floor annex.[7] Cameras were installed in the elevator lobby on each floor and at the entrance of the building.

The University of Georgia Libraries are still in the process of recovering from the fire, and will be for quite some time. The binding has been completed and the items that could not be bound have been placed into envelopes and returned to the stacks. The Libraries' federal depository status has assisted greatly in reacquiring many of the missing items from other depository libraries weeding their collections, and some of the items have been purchased again, thanks to insurance money. However, the effort of processing these replacements, as well as processing materials that had been sent for rebinding, is a time intensive task and, while the work with the rebinds is almost complete, the task of finding and collecting replacement items could go on forever. In addition, those items that survived the fire have been exposed to extreme environmental conditions, greatly reducing their durability and their shelf-life.

WHAT WAS LEARNED?

The lessons that the preservation team learned from this major disaster and from the recovery efforts fall into two rather broad categories: lessons learned about the nature of fire damage and lessons learned about the disaster plans. As mentioned earlier, the preservation literature dealing with fire-damaged materials is scant, and the governing assumption appears to be that if material is burnt, it should simply be thrown away.[8] The University of Georgia Libraries fire suggests, however, that material that has been exposed to fire can often survive and be made useable once again, for a fraction of the price of replacing material. For instance, rather than replace large portions of the U.S. Congressional Serial Set, which would have been very costly, these items could instead be rebound for less than ten dollars an item, with no real effect on the usability of the item. Rebinding, then, was a major possibility for a majority of the bound materials that were directly affected by the fire, as well as a large number of paperbound materials that had not sustained excessive damage. These items were sent to a professional bindery company, the burnt casings were disposed of, the burnt edges were trimmed, and the items were placed in new bindings, leaving them in a state similar to their condition before the fire.

Also, all of the items in the rest of the building that had sustained soot damage, as well as many of the items on the second floor that sustained the heavier soot damage, could simply be cleaned. If the University of Georgia Libraries had decided to dispose of the material that had simply suffered soot damage, the majority of its collections would have been destroyed. The possibility of saving a large portion of the collection by rebinding or cleaning created a situation in which the number of items destroyed by the fire could be radically decreased, but it also created a situation in which the number of items personally handled by library staff increased drastically, introducing the element of human error into the recovery efforts.

The most obvious error that the recovery team committed had to do with the conditions of the environment in the fire-damaged area and the kind of damage the fire caused. The area in which the damaged materials were contained, after many of them were moved, had extremely poor lighting, necessitating that individual staff members spent a long period of time studying items to determine whether or not they could still be used, and complicating the ability to make that determination. It is very difficult indeed to determine if a substantial portion of the content of an item is intact when one cannot properly see the content at all. The difficulty posed by the environmental conditions meant that some items that contained maps or other kinds of foldouts were sent to be rebound; this caused damage when the items were trimmed during the binding process. In addition, even with items that retained their content, the fire burnt the casing of most items directly affected by it, and therefore the call numbers on those items. In cases where the call numbers were burnt, but the items were intact, a government documents staff member would have to attempt to determine the call number. It was impossible in most cases where the items were completely burnt to determine a call number at all, causing the library staff to determine the call number that preceded that range of burnt items and the call number at the end of that range and declaring everything between those call numbers to be destroyed.

However, because items had been moved out of their original location in the stacks, this produced an inflated number of destroyed items that then had to be revised as the recovery process continued. As the material came back from the bindery, and as the government documents staff was able to work through the material that had been cleaned and returned, the staff determined what items had been destroyed based on what was not accounted for but that the University of Georgia Libraries certainly owned. The process of determining what had been destroyed was obviously a complicated one, but it was a necessary step so that the University of Georgia Libraries could account for what had been destroyed for insurances purposes as well as to determine what needed to be replaced and to update the cataloging records.

One issue associated with the fire that has not been discussed thus far, but of concern during the recovery process, is the smell of the fire-damaged material. Simply put, nothing can really be done about the smell of burnt materials. The usual methods for combating this problem, ozone treatment or thermal fogging, are completely inappropriate, since they will do additional harm to the materials that have been already exposed to damaging environmental conditions. The University of Georgia Libraries brought in machines that emitted an agent designed to neutralize the source of the smoke odor. This solution worked to a great extent. At present, there is a slightly burnt odor near the materials that were heavily affected by the fire, and certainly if one were to use these materials the scent would be even more pronounced, though certainly less noisome than it could have been. The odor, however, is fairly mild, and even if it stays with the material for quite some time, it does not affect the usability of these items.

Finally, the library staff learned certain important lessons regarding the nature of disaster plans. The University of Georgia Libraries' disaster plan did not turn out to be all that useful. There were a few circumstances that contributed to this situation. First, the disaster plan had not been updated in some time, and many of the important contact numbers were not on it. The fact that the disaster plan was outdated was not the most significant problem with it, because, secondly, no one had a paper copy of the plan; once it was housed on the University of Georgia Libraries' server, there was a general assumption that paper copies were obsolete. However, once the building was on fire, the servers were shut down in order to secure them, meaning that the disaster plan was not accessible. Third, the disaster plan did not detail many of the issues that library staff experienced in contacting and working with a disaster recovery company; the disaster plan had simply informed the staff to hire such a company without providing the staff with a blueprint on how to work with such an organization.[9]

The facts that the disaster plan was outdated, and that it was inaccessible for the initial part of the recovery, were not as significant an obstacle to the fire recovery efforts as one might expect, because the staff worked together well in the absence of the disaster recovery plan. The staff found that many of the steps which disaster plans spend quite a bit of space detailing occurred almost naturally. Setting aside material to triage, for instance, was the only logical thing to do in this situation. It was clear that the preservation unit needed to be involved and that they would need to work with the staff members who were responsible for the part of the collection most directly affected. After the preservation unit had worked very briefly with both the disaster recovery company and the department responsible for the collection at large, the stacks supervisor in this case, the recovery efforts could be carried out by other individuals leaving the preservation unit free to work with the area in which they were needed the most.

In addition, both of the units involved in these efforts worked closely with the University of Georgia Libraries' director, who acted as a liaison between the recovery teams and the insurance companies, making certain that the recovery teams received the assistance required as soon as they possibly could. In short then, all the disaster plan would need to point out is that the preservation unit needs to quickly form a working unit with the staff directly responsible for the affected area so that these units can make decisions in an efficient manner. It is not that disaster plans are unnecessary, but, as this fire demonstrates, disasters are unique. Disaster plans could not possibly cover all of the possible scenarios that an institution could encounter in a major disaster. Instead of attempting to discuss all of those possible scenarios, disaster plans should detail how to bring together different library units, each with training appropriate to the specific disaster.

A disaster recovery plan could have been more useful if it gave the library staff some sort of expectation about what working with a disaster recovery company would be like. In this disaster, the disaster recovery company was absolutely necessary. The company was able to acquire massive quantities of materials overnight, materials such as dry chemical sponges, paper towels, industrial humidifiers, and industrial fans, both to control the environment of the affected area (something the staff had neither the ability nor experience to do) and to clean the lesser-affected material. In addition, the disaster recovery company was able to offer significantly more manpower than the University of Georgia Libraries could have done.

There were a few instances, however, where this relationship was more problematic. Disaster recovery companies are more familiar with handling disasters in which large amounts of the affected materials could be disposed of, or in which the material has no artifactual value. As such, these companies might offer to use ozone or thermal fogging to alleviate some problems. These options are not suitable for research libraries, and the staff needs to be able to reject these offers, as the preservation unit did.

Disaster recovery companies are likewise unfamiliar with library organizational systems, and while the disaster recovery company was able to clean the materials affected by the fire throughout the building, the employees often reshelved those items out of order within a shelf. Once the building reopened and the library staff realized that the items throughout the building were misshelved, the staff had to initiate a library-wide shelf-reading and reorganization project. These situations were unexpected and a disaster plan that spends more time discussing how one should work with disaster recovery companies would be useful.

It is our hope that the experience of this fire will be able to help other libraries in instances where they have experienced some form of disaster and in instances where they are modifying their disaster recovery plans. Fire damage, as the University of Georgia Libraries experienced it, was not as problematic as we had been led to believe. Rebinding and cleaning are viable options for most affected items and the damage, in some ways, is less problematic than water damage, since the treated material often shows no signs of ever being in a disaster. In addition, disaster plans that spend too much time detailing what should be done to treat materials, when the disaster is small enough to be handled in-house, are not as useful as they could be. These plans should state that certain units must be brought together and be allowed a certain amount of flexibility in the recovery decision-making process. The plans should spend more time discussing in depth how to contact and work with outside help from a wide variety of companies, such as binderies and disaster recovery companies, giving library employees some idea of what to expect and the ways in which these companies can most effectively help their disaster recovery efforts.

NOTES

1. The Athens-Clarke County Police Department determined that the fire was caused by arson. As of this writing one individual has been arrested, tried, and acquitted in April of 2005 for setting this fire.

2. The major work in this field, and instructive as to the dearth of information of fire damage as opposed to water damage, is Sherelyn Ogden, ed, *Preservation of Library & Archival Materials: A Manual* (Andover: Northeast Document Conservation Center, 1999).

3. For a few sample plans of this sort, which likewise present a more detailed concern with water damage as opposed to fire damage, see the following: University of Indiana, "Disaster Response Plans," University of Indiana, <http://www.indiana.edu/~libpres/disaster/baseplan.html> (accessed August 1, 2006); Harvard University, "Emergency Preparedness Plan: Templates," Harvard University, <http://preserve.harvard.edu/emergencies/plan.html> (accessed August 1, 2006); New York University, "Disaster Plan Workbook," New York University, <http://library.nyu.edu/preservation/disaster/toc.htm> (accessed August 1, 2006).

4. The items were not allowed to sit intentionally. The second floor was originally considered a crime scene, since the fire was arson, and therefore the staff could not get to it right away. I will return to this point later.

5. Because they were concerned and eager to assist, and because they could not get back to their normal duties in most cases, the staff volunteered readily, allowing the preservation team to build a relatively sizable recovery team.

6. These sponges are made chemically; they do not have chemical content.

7. A sprinkler system is in the process of being installed in the rest of the building.

8. Judith Fortson, *Disaster Planning and Recovery: A How-To Manual for Librarians and Archivists* (New York: Neal-Schuman Publishers, 1992) makes the usual move of claiming that "even with the more obvious calamity of fire, actual damage often comes from water used for suppression" (52). In the University of Georgia fire, this was simply

not the case. *Preservation of Library & Archival Materials: a Manual* likewise seems to assume that the main concern after a fire is water damage, only if the "collections survive at all" (126).

9. This oversight is true of many disaster recovery plans. See, for instance, the plans listed in note three.

doi:10.1300/J118v25n03_07

Steamy Situation: Water Emergency in Sterling Memorial Library

Tara D. Kennedy

SUMMARY. During construction in an adjacent building, a steam pipe burst in Sterling Memorial Library at Yale University, causing the fire sprinklers to discharge in the basement of the library, affecting a backlog of collections, including important newspapers from Cambodia. Twenty-three library personnel, along with an outside firm, Munters Moisture Control, were able to recover the majority of the library materials with minimal loss. doi:10.1300/J118v25n03_08 *[Article copies available for a fee from The Haworth Document Delivery Service: 1-800-HAWORTH. E-mail address: <docdelivery@haworthpress.com> Website: <http://www.HaworthPress.com> © 2006 by The Haworth Press, Inc. All rights reserved.]*

KEYWORDS. Vacuum freeze-drying, construction, steam pipe, emergency preparedness, disaster recovery, emergency plan, Post-it Notes™, newspapers, Cambodia, architectural drawings

---

Tara D. Kennedy, MLIS, is Preservation Field Services Librarian, Yale University Library, Preservation Department, P.O. Box 208240, New Haven, CT 06520-8240 (E-mail: tara.d.kennedy@yale.edu).

[Haworth co-indexing entry note]: "Steamy Situation: Water Emergency in Sterling Memorial Library." Kennedy, Tara D. Co-published simultaneously in *Public Library Quarterly* (The Haworth Information Press, an imprint of The Haworth Press, Inc.) Vol. 25, No. 3/4, 2006, pp. 89-97; and: *Dealing with Natural Disasters in Libraries* (ed: William Miller and Rita M. Pellen) The Haworth Information Press, an imprint of The Haworth Press, Inc., 2006, pp. 89-97. Single or multiple copies of this article are available for a fee from The Haworth Document Delivery Service [1-800-HAWORTH, 9:00 a.m. - 5:00 p.m. (EST). E-mail address: docdelivery@haworthpress.com].

Available online at http://plq.haworthpress.com

doi:10.1300/J118v25n03_08

INTRODUCTION

The Law of Water Emergencies in Libraries is as old as the institution itself: water disasters will happen on the weekend, over or near a holiday recess, and most likely because of construction. Sterling Memorial Library at Yale University was no exception to this law; on Saturday, January 7, 2006, a steam pipe burst in an adjacent student residence under renovation, Trumbull College, forcing steam into the library basement. The excessive heat set off four sprinkler heads in the south side, lower level of the library, soaking bound and unbound collections in Circulation, the Manuscripts and Archives collection, the Slavic Reading Room, and one of the Southeast Asia and Near East catalog backlog areas (Illustration 1).

The steam-soaked ceiling tiles and condensation on ceilings and pipes caused the rest of the water damage (Illustration 2).

Facilities and library emergency plans, and the emergency call list went into effect, notifying managerial-level staff in the Preservation Department and all department heads with damaged collections. The

ILLUSTRATION 1. Water Damage to Asian Collections

Reprinted with permission.

ILLUSTRATION 2. Damaged Ceiling Tiles

Reprinted with permission.

Yale University Emergency Operations Plan did not go into effect, as this water emergency rated as a moderate event, which did not require the assistance of additional university personnel or supplies. However, twenty-three staff members from the library gave up their weekend to assist in the recovery of materials.

RESPONSE EFFORTS

Initially, library staff members started the recovery effort by drying materials in available places in Sterling Memorial Library. Preservation Department staff members quickly instructed library personnel on how to handle the wet materials and fan out bound materials to air dry. For the severely soaked materials, the library was fortunate in that Yale University Risk Management had the foresight to have Munters Moisture Control Services under contract with the university. Munters Moisture Control Services is a division of the Munters Corporation that specializes in emergency recovery of fire and water-damaged documents and media. Munters' personnel worked quickly to reduce the risk of mold in the library, as saturated collections either were removed quickly from the humid environment into freezer trucks, or were dried partially *in situ* through the dehumidification equipment used by Munters, such as industrial fans and dehumidifiers. Munters workers labored over the weekend to reduce the relative humidity in the basement areas and to remove wet materials for vacuum freeze-drying treatment. Vacuum freeze-drying is a process that removes water through sublimation–going from a solid phase directly into a gas phase–under a strong vacuum. It is suitable for many paper-based materials, including newspapers and books (especially books with coated paper), which were the majority of materials that were damaged during this water emergency. It is not suitable for any type of photographic print or negative, vellum bindings, magnetic media, or computer media.

The majority of materials sent to be vacuum freeze-dried were from an important section of the Southeast Asian collection. Yale University Library has played an important role in collecting evidence of genocide by the Khmer Rouge through microfilming the holdings of the Documentation Center of Cambodia. This large collection of 482 reels depicts records of the Khmer Rouge-era Santebal (secret police) files and other related records. Thankfully, the steam pipe break did not affect the microfilm reels, but 30 boxes containing 58 newspaper titles from Cambodia, collected during the United Nations Transitional Period in Cambodia (and intended for this same microfilming project), were saturated by the sprinkler system (Illustration 3).

This period is of particular interest to those studying Cambodian history, as it documents the significant period of the redevelopment of free speech and journalism after the long, oppressive Khmer Rouge rule, followed by war between the Khmer Rouge and Vietnam (circa 1989-1995). Yale University Library is the unique holder of many of these titles. In addition to the Cambodian newspapers, approximately 4,500 monographs and serials in the cataloging backlog, mostly from the Southeast Asian collection, were damaged from the activated sprinkler system and condensation from the ceiling and pipes. An interesting problem occurred with some of the newspapers that had gotten wet during this emergency. Post-it Notes™ and similar "removable" adhesive notes on these newspapers had gotten wet and bled a pink color onto the objects (Illustration 4). This situation is just another reason why these notes should not be used on objects.

ILLUSTRATION 3. Damaged Cambodian Newspapers

Reprinted with permission.

Some of the materials were affected only by the condensation "rain" and were recovered in-house (Illustration 5). One such collection was from the Manuscripts and Archives department, the Eero Saarinen Architectural Drawing collection. This collection was sitting in a processing area in paper folders. Through rapid action, the drawings with wet folders were placed in new folders in order to prevent the water from penetrating onto the drawings. This quick thinking prevented further damage to the majority of the collection in this area. Placing damp drawings between blotter papers dried other drawings that had become saturated by the condensation. Preservation staff frequently changed the blotter paper, as often as every half-hour in some cases, to ensure that the items dried quickly with as little distortion as possible.

Beyond collections, the flood caused problems for the library's computer networking system. Four library computer servers were located in a secure room in the library's basement. The high temperatures from the steam caused the servers to turn off and the condensation from the ceiling and pipes made the servers wet. Staff of the library's Web, Workstation,

ILLUSTRATION 4. Ink Bleeding from Sticky Notes

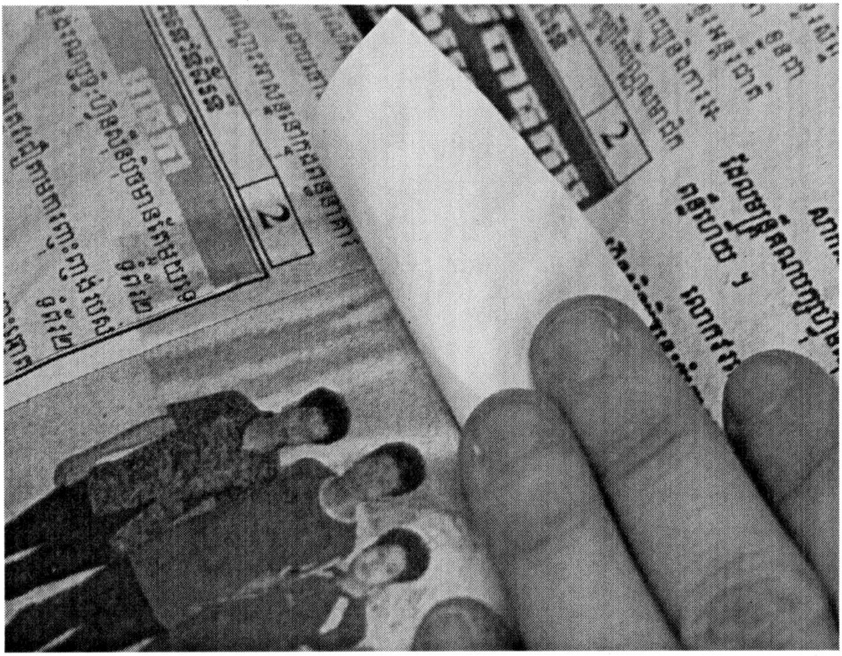

Reprinted with permission.

and Digital Consulting Services group immediately placed their emergency downtime plan into effect. Services were transferred to an offsite backup system within 15 minutes of being notified of the water emergency at the library. The servers had to be replaced due to the water and heat damage.

Thanks to the prompt response from Munters and library staff, only 46 of the volumes that were vacuum freeze-dried and/or air dried in-house needed to be replaced, an astonishing recovery rate of 98.97%. Most of the volumes that needed to be replaced were because of "blocking," a description of what occurs when coated paper becomes wet and fuses together; staff had not reached these items in time to prevent the coated pages from sticking together. The newspaper recovery rate was even higher than the books: not a single issue was lost because of the water emergency, with only minor undulations in the newspapers (Illustration 6).

To what do we owe this high percentage of recovery for these materials? Firstly, the prompt response of the library staff in all departments.

ILLUSTRATION 5. In-House Recovery Efforts

Reprinted with permission.

While no emergency goes smoothly, all persons who needed to be contacted were notified in a timely manner, from the Buildings Operations Manager to the Head of Preservation to the University Librarian. Having an up-to-date list of contact information as a part of the library emergency plan expedited this process. Excellent communication on all levels during the emergency also assisted with the success of this situation. The Library Emergency Plan outlines the responsibilities of the individuals who would be called first in an emergency, specifically the Preservation Department and Buildings Operations and Security. In the emergency plan, the Preservation Duty Officer (the Preservation Department manager in charge during the emergency) and the Buildings Operations Manager are the Preservation Coordinator and the Facilities Coordinator, respectively. Each of these positions has a list of responsibilities; the Preservation Coordinator is accountable for:

- The assessment and reporting of the emergency situation concerning collections

ILLUSTRATION 6. Undamaged Newspapers

Reprinted with permission.

- Ensuring the protection of undamaged collections
- Assessing extent and nature of damage to collections
- Recommending and performing/instructing actions and treatment regarding collections
- Communication to other administrative officials at the university concerning collections including Risk Management, Office of the Secretary, the University Librarian, library staff, Library Business Office, and the Workstation Support Specialist
- Documentation of collections-related decisions and actions, including recovery
- Welfare of staff performing recovery work by ensuring breaks are scheduled, for example

The Facilities Coordinator has the following duties during an emergency:

- Assessment and reporting of the emergency situation in the facility
- Ensuring control of the emergency in the facilities, shutting off water, contacting service vendors, for example

- Ensuring protection of undamaged portion of facility
- Assessing extent and nature of damage to facility
- Communication to other administrative officials in the university concerning the facility including Risk Management, Office of the Secretary, the University Librarian, library staff, Library Business Office, and the Workstation Support Specialist
- Contacting Local Building Operations Manager and Security or Evening/Weekend Security Supervisor, as needed, including contacting the appropriate Associate University Librarian, as needed
- Liaisons with civil authorities (if necessary)
- Facility clean up and recovery efforts
- Security arrangements
- Documentation of facilities-related decisions and actions

Knowing these assignments before an emergency allowed for better time efficiency. Available resources were also an essential benefit; if not for the existing vendor agreement with Munters and the University, this recovery effort would have been far more strenuous and demanding on library staff, especially Preservation and Southeast Asia Collection staff.

LESSONS LEARNED

Most of the damage was a result of the sprinkler system, which released water when the room reached a certain temperature. One remedy to that situation: the sprinkler heads were replaced and set to a higher temperature threshold of 165°F. This incident also prompted the library to move the servers to a less vulnerable area, outside of the basement and Sterling Memorial Library. Curators of collections have taken better precautions in protecting their collections in this area. An example of this is the protection that the Cambodian newspapers now receive; the newspapers are in acid-free boxes, on palettes, and covered with plastic drop cloths.

doi:10.1300/J118v25n03_08

Riding the Surf: Dealing with Library Disasters in Island Communities

Lynn Ann Davis

SUMMARY. To respond effectively to a disaster requires knowledge, flexibility, planning and practice. Developing a sustainable model in Hawaii and the Pacific region for disaster planning and preservation is the logical extension of the lessons learned from the flash flood in October 2004 at the University of Hawaii at Manoa (UHM) Library. Surfing is a useful metaphor. doi:10.1300/J118v25n03_09 *[Article copies available for a fee from The Haworth Document Delivery Service: 1-800-HAWORTH. E-mail address: <docdelivery@haworthpress.com> Website: <http://www.HaworthPress.com> © 2006 by The Haworth Press, Inc. All rights reserved.]*

KEYWORDS. Disaster response, disaster planning, maps, aerial photographs, Hawaii, Pacific region

---

Lynn Ann Davis, MLS, is Head of the Preservation Department, University of Hawaii at Manoa, Hamilton Library, 2550 McCarthy Mall, Honolulu, HI 96822 (E-mail: ldavis@hawaii.edu).

The photos in this article are courtesy of Susan Murata and Julie Page. Reprinted with permission.

[Haworth co-indexing entry note]: "Riding the Surf: Dealing with Library Disasters in Island Communities." Davis, Lynn Ann. Co-published simultaneously in *Public Library Quarterly* (The Haworth Information Press, an imprint of The Haworth Press, Inc.) Vol. 25, No. 3/4, 2006, pp. 99-112; and: *Dealing with Natural Disasters in Libraries* (ed: William Miller and Rita M. Pellen) The Haworth Information Press, an imprint of The Haworth Press, Inc., 2006, pp. 99-112. Single or multiple copies of this article are available for a fee from The Haworth Document Delivery Service [1-800-HAWORTH, 9:00 a.m. - 5:00 p.m. (EST). E-mail address: docdelivery@haworthpress.com].

Available online at http://plq.haworthpress.com

doi:10.1300/J118v25n03_09

INTRODUCTION

Surrounded by ocean, and waves breaking on reefs and beaches, islands are separated from each other and from continents. Isolation compounds the problems faced by librarians in responding to disasters: flooding in tropical environments requires immediate action to prevent mold growth; island communities rarely have preservation professionals or conservators on-site to lead recovery efforts; and resources take days, even months, to arrive on islands. It is essential, therefore that librarians, living in island communities, are prepared to respond effectively to a disaster.

Preparedness and ocean safety are taught to island children before they attend school. Children not only learn swimming and ocean sports, but also are drilled on essential safety rules such as never turning their backs to the surf; always being aware of what is happening in the ocean; and knowing local conditions and currents. This level of alertness can be incorporated into disaster training and preparedness for libraries and other cultural institutions.

There are six main islands that compose the Hawaiian Islands. Public libraries are part of a state supported education network. Each island community is served by at least one main public library. The University of Hawaii system has campuses and libraries on four of the islands (Kauai, Oahu, Maui, and Hawaii Island). The University of Hawaii at Manoa Library (UHM), located in Honolulu, is the research library for the state and the University system. In October 2004, the UHM Library faced a major disaster caused by torrential rain.

It seems appropriate to use surfing as a metaphor for how an island librarian could handle a disaster. Surfers use skill, knowledge of local surf conditions, and timing to catch a wave. After sizing up a wave, they paddle ahead of the on-coming swell, and with a burst of power, they drop over the lip of the wave just as it starts to break. Ideally a well-designed disaster preparedness plan would function like riding the surf, preparing librarians to be able to respond quickly and effectively to any disaster that threatened collections.

Surfing is also a personal metaphor. As a body surfer, there is that special moment as you are swimming out to catch the next wave, when you confront an unexpected set of waves. By diving deep under the wave, it roars over you, sounding like a passing train, and suspends you for a brief moment in an incredible stillness. Leading a large-scale disaster recovery requires making quick decisions, and responding to several crises at once. The disaster, like confronting the breaking wave, can be overwhelming unless you are able to find moments of quiet, to unravel and resolve the vast array of problems that require immediate attention. The set of breaking waves, initiated by a disaster, continues for years.

A flash flood swept down Manoa valley and hit the University of Hawaii at 8 p.m. on Saturday, October 30, 2004. In the Library it created a powerful set of waves surging up to nine feet on the ground floor, throwing desks, book trucks, books, boxes, map cases, shelving, and files against the building's supporting walls. The floodwater carried mud and debris through the campus. In the Library the first responders assessed the damage with flashlights, moving through a sea of mud that had been deposited around the building. The fecund smell of the mud became a part of the collective memory of the disaster salvage and recovery.

The Library's collection plan for disaster recovery can be summed up in four words: *act before mold grows*. Collection disaster recovery is coordinated by the Preservation Department; the plan is on the web and in library departmental manuals; and a broad spectrum of staff has been trained in disaster response techniques. Previously the Preservation Department staff had successfully responded to smaller disasters (up to 1,000 books) caused by construction. The scale of this disaster, however, required ramping up the existing plan, being creative, flexible, and relentless in achieving the goal of stabilizing the collections.

The ground floor of the UHM Library housed the Acquisitions, Serials, and Cataloging departments where materials were being processed. In this area, shelves were thrown into heaps with desks, chairs, books, etc. (Illustration 1).

The Library's computer server was destroyed, and tadpoles were found swimming in the flooded room. Students attending a class in the graduate program of Library and Information Sciences broke a window to escape the power of the surging water. The Library's salvage efforts were focused on the Map Collection, a part of the Government Documents Collection, also located on the ground floor (Illustration 2).

The disaster response had two important stages. Based on significance, and uniqueness, the Hawaii and Pacific region maps and aerial photographs were the first priority to stabilize. With the assistance of the library staff and volunteers, map drawers were moved directly into freezer containers (Illustration 3). All the priority materials (over 60,000 maps and 70,000 aerial photographs) were frozen within 72 hours. Books from the government document collection that could be salvaged were also frozen. Card catalogs for the government document collection, aerial photograph collection, and the shelf list for the Library's Asia vernacular collection were located and frozen.

ILLUSTRATION 1. Entrance to the Cataloging Department at UHM

By immediately freezing the priority materials, we not only prevented mold growth, but we also took charge of the recovery and treatment by buying time that allowed us to make decisions and develop a protocol for treatment of the materials (Illustration 4). Much as a surfer comes up for air and then dives under the next on-coming wave, freezing materials provided a chance to catch our breath.

The second stage of the disaster response also required immediate action. The temperature and relative humidity in the Library building had to be controlled to prevent mold growth on books. To achieve this goal, temporary air ducts were positioned in the windows of the five floors above the flood damaged ground floor (Illustration 5); translucent tubes snaked throughout the building and fans distributed the cooled, dry air. The Library operated using several small generators, and lights were strung throughout the stacks which gave an appearance of a circus midway. Although it took several months to bring in (by ship) a generator that was large enough to power the Library's electrical needs, the Library managed to open within a month by paging books, and the Library web page and electronic resources were up within a week.

ILLUSTRATION 2. The Map Collection Room Immediately After the Disaster

With a disaster of this scale, disaster recovery companies converged on the University and the Library seeking to get contracts. One American colleague characterized them as carpetbaggers. In the Pacific region "experts" come and go, receiving large fees, and implementing changes without being interested in the local perspective. Since the materials were frozen, the Library could take time to determine the services we wanted and identify a company that was the best match for our needs. Communication and experience in library document recovery were the most important criteria. Trust was also essential since we would be working across several time zones and thousands of miles with any American company. In the initial discussions with Belfor U.S.A., they did not assume that they knew more than the local professional staff, and were interested in working together to develop the protocol for treatment of mud and water damaged maps (Illustration 6).

Every disaster response/recovery has its own ebb and flow. An informal network of conservators and preservation professionals provided information, and emotional support over the phone; Library Directors at Brigham Young University at Provo, Utah, and University of California

ILLUSTRATION 3. Library Staff Carrying Map Drawers to the Freezer

San Diego sent professional staff out within the first two weeks to help with our recovery efforts. In January 2005, Belfor staff worked with our staff and a group of conservation and preservation consultants from America and New Zealand to examine damage and establish treatment protocols.[1] Sample drawers of the maps were removed from the freezer for examination. The force of the water penetrated all the map drawers, and the maps were coated with mud. Damage was documented and treatment procedures were discussed and tested (Illustration 7).

Permanent damage (a purple stain) was caused to maps by the unstable dye used in the cloth covers found in many of the map drawers (Illustration 8); ink from property stamps transferred to other maps, and identification written in ink bled and stained the maps (Illustration 9).

The University of Hawaii Administration's up-front fiscal support for the Library's initiatives was a major factor in the successful disaster recovery effort. Three months after the flood, the University was declared a federal disaster, enabling FEMA to become involved in February 2005. However, it will be years before the University is reimbursed by FEMA through the Hawaii State Department of Civil Defense. Leslie Paisley, a paper conservator, was assigned by FEMA to work with the Preservation Department. A forty-page protocol, approved by

ILLUSTRATION 4. Maps, Aerial Photographs, and Other Library Materials were Stabilized in Five Freezer Containers

FEMA, outlines three levels of treatment for unique and rare maps of Hawaii that have not been shipped to Belfor. The treatment goal for the first two levels is to return the damaged maps to a "function level" so that they can be used by researchers. A paper conservator is necessary to complete additional treatment for rare maps to restore them for exhibition.

After being rinsed to remove the mud and freeze-dried by Belfor in Fort Worth, Texas, 50,000 maps have returned to Hawaii. A forty-foot shipping container of wet books from the government document collection has been freeze-dried and returned; and the Asia vernacular catalog was dried and scanned for future reference.

Libraries in Hawaii and the Pacific region generally do not have facilities and trained staff for conservation treatment for collections. UHM Library's Preservation Department's lab (built in 1999) was designed to accommodate the Library's future needs, as well as the lack of conservation facilities in Hawaii. The disaster had a silver lining; resources were available to equip the existing paper conservation and the digital lab, and develop the skills of the local staff through training and on-site

ILLUSTRATION 5. External Air Ducts Delivering Cool, Dry Air to Upper Floors of Hamilton Library

apprenticeships with consulting paper conservators. Training empowers the staff, and creates a new sustainable level of expertise in basic paper conservation (Illustration 10) and digital documentation.

Two years after the disaster the Preservation Department continues to work on recovery of Library materials. Approximately 8,000 maps and other materials have been treated in the UHM Library Preservation

ILLUSTRATION 6. Rachelle Helms Training Staff to Rinse Maps at Belfor, Fort Worth, Texas

Department (including rare maps) out of 10,000; 2,000 remain to be treated in the UHM Library Preservation Department Lab.

During the spring semester of 2007, the department will host a symposium on encapsulation to discover what happened to cause the mylar welds to break when the flood water slammed against the map cases (Illustration 11).

The research into these issues will be led by William Minter, paper conservator and developer of the ultra-sonic welder for encapsulation used in the Library. Potential causes will be examined, and UHM procedures for encapsulation will be reviewed to insure that adequate standards for training staff and student workers will be implemented in the future.

Treatment of 70,000 aerial photographs of the Western Pacific region is the final challenge. The photographs were made after WWII, and are printed on a variety of B&W photographic papers. In April 2007, Gary Albright, a well-known photograph conservator, worked at UHM Library with the staff and other consultants to examine the damaged photographs and develop a protocol for treatment. Since the photos were

ILLUSTRATION 7. Consultants Julie Page and Mary Wood Lee, Documenting Damaged Maps in UHM Library Preservation Department, January 2005

tightly packed in file drawers it is possible that most of the damage caused by mud will be confined primarily to the edges.

A little more than a year after the UHM Library disaster, another wave hit the shores of islands in the Pacific. The release of the seminal report by Heritage Preservation, *A Public Trust at Risk: Health Index Report on the State of America's Collections,* noted that in the Western region of the U.S. (which includes American Samoa, Commonwealth of the Northern Marianas Islands, Guam, and Hawaii) 81% of the institutions surveyed do not have "emergency plans with staff trained to carry it out."[2]

Despite this demonstrated need, it is difficult for institutions to make time to focus on disaster planning and staff training in the midst of other pressing demands. To get beyond the understandable resistance, there needs to be a fresh, sustainable, and flexible approach to disaster planning and training. A group of preservation professionals, who formed the Western States & Territories Preservation Assistance Service (WESTPAS),[3] received a National Endowment for the Humanities (NEH) grant to assist

ILLUSTRATION 8. Permanent Staining Caused by Color Bleeding from Cloth Cover in Map Drawer

ILLUSTRATION 9. Ink Marking Bleeding to the Front of a Map

ILLUSTRATION 10. Kyle Hamada, Paper Conservation Technician, Rinsing Map in the UHM Library Preservation Department Lab

libraries and archives in the Western region of the United States to develop disaster response plans, strengthen professionalism, and create a collaborative network. This sustainable model for disaster recovery, supported by librarians and archivists in the region, develops professionals by expanding their knowledge and experience of preservation practices. It also creates an opportunity for collaboration by sustaining a virtual network for sharing information through the internet, and informal networks that provide assistance in disaster recovery.

Trained professionals will present two separate workshops at each location. The first will "deliver preservation education and training on disaster preparedness, emergency response, and collection salvage,"[4] and work with participating institutions to develop disaster response plans that reflect their local circumstances. The follow up visit is to insure that participating institutions have devoted the staff time to develop a disaster plan. It is essential to have the disaster plan in place, to train staff, and build a flexible approach to disaster response.

ILLUSTRATION 11. Wet Map Expanded Within Mylar Encapsulation Causing the Paper to Cockle

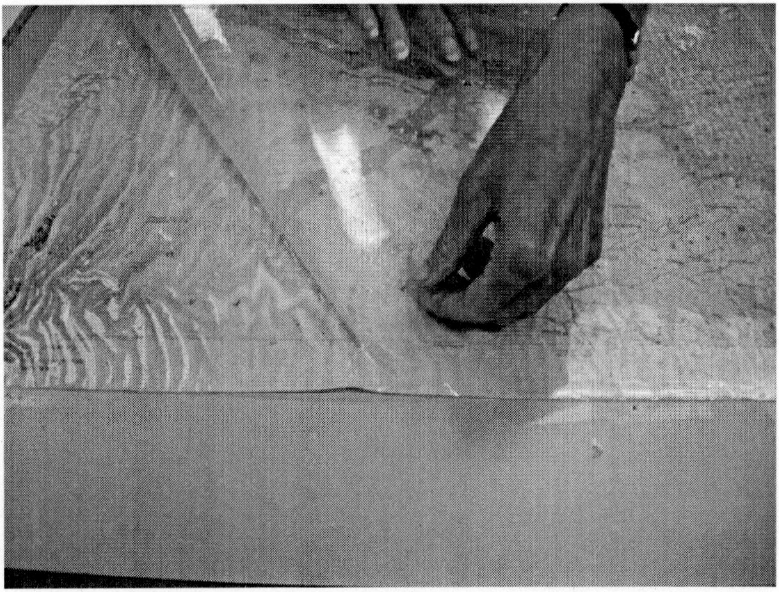

Surfers are familiar with how storms, winds and currents affect the waves. By observing the waves from the beach, surfers apply their experience, and develop a flexible plan to catch the ultimate wave. As professionals, we know potential disasters may damage our collections. With experience and training, librarians, archivists and museum professionals will be able to approach any disaster with the same confidence of a skilled surfer preparing to catch a wave.

NOTES

1. Consultants to UHM Library Preservation Department: Mary Wood Lee (paper conservator, Connecticut), Randy Silverman (Marriot Library, University of Utah), Julie Page (University of San Diego), Pam Najar-Simpson (National Library of New Zealand), Graham Simpson (photo conservator, New Zealand), and the Belfor staff, including Kirk Lively and Rachelle Helms.

2. Heritage Preservation, *Heritage Health Index Full Report*, "Chapter 7: Emergency Planning and Security," Figure 7.5, p. 63 <http://www.heritagepreservation.org/HHI/HHIchp7.pdf>, accessed January 15, 2007.

3. The program planning group convened in March 2006 by Barclay Ogden (Head of the Preservation Department, University of California at Berkeley) included: Lynn Davis (University of Hawaii), Normandy Helmer (University of Oregon), Kristen Kern (Portland State University), Gary Menges (University of Washington), Julie Page (University of California San Diego), Terri Schindel (Museum Training Network, Colorado), and Randy Silverman (University of Utah).

4. Email attachment from Barclay Ogden to Julie Page and Lynn Davis, December 12, 2006.

BIBLIOGRAPHY

Heritage Preservation, *A Public Trust at Risk: Health Index Report on the State of America's Collections,* <http://www.heritagepreservation.org/HHI/index.html>.

Heritage Preservation, *Case Study*, "Hamilton Library (University of Hawaii at Manoa)," <http://www.heritagepreservation.org/HHI/case91.html>, accessed January 15, 2007.

Moreno, Loren, "Library Maps Returning to View," *The Honolulu Advertiser*, January 8, 2007, p. 1.

"Save Time and Your Collection: Be Prepared When Disaster Strikes," *Western Trek: The OCLC Western Magazine*, vol. 2, no. 1, Fall 2006 p. 2-3.

doi:10.1300/J118v25n03_09

The Complicated Symptoms of Sick Buildings

Richard P. Widdicombe

SUMMARY. When the symptoms of a possible sick building syndrome appeared in a library building more than twenty years ago, it was thought that the situation could be easily fixed. Instead, we are still managing a problem building although extensive work has been done and much progress has been made to moderate the physical reactions of staff and building users. Looking after the well being of our staff has been the primary motivation. doi:10.1300/J118v25n03_10 *[Article copies available for a fee from The Haworth Document Delivery Service: 1-800-HAWORTH. E-mail address: <docdelivery@haworthpress.com> Website: <http://www.HaworthPress.com> © 2006 by The Haworth Press, Inc. All rights reserved.]*

KEYWORDS. Sick Building Syndrome, SBS, HVAC systems, interstitial spaces, reactions to environment, mold, employee symptoms, managing illness in employees

---

Richard P. Widdicombe, MSLS, is Frederick Winslow Taylor Curator at Stevens Institute of Technology, 1711 Sherman Dr., Utica, NY 13501 (E-mail: richard.widdicombe@gmail.com).

[Haworth co-indexing entry note]: "The Complicated Symptoms of Sick Buildings." Widdicombe, Richard P. Co-published simultaneously in *Public Library Quarterly* (The Haworth Information Press, an imprint of The Haworth Press, Inc.) Vol. 25, No. 3/4, 2006, pp. 113-125; and: *Dealing with Natural Disasters in Libraries* (ed: William Miller and Rita M. Pellen) The Haworth Information Press, an imprint of The Haworth Press, Inc., 2006, pp. 113-125. Single or multiple copies of this article are available for a fee from The Haworth Document Delivery Service [1-800-HAWORTH, 9:00 a.m. - 5:00 p.m. (EST). E-mail address: docdelivery@haworthpress.com].

Available online at http://plq.haworthpress.com

doi:10.1300/J118v25n03_10

INTRODUCTION

When maintenance questions your sanity and doctors treat just symptoms, how do you manage a sick library building? In dealing with this problem for almost twenty years, I have often thought of the poem "The Blind Men and the Elephant" by John Godfrey Saxe (1816-1887):

It was six men of Indostan,
To learning much inclined,
Who went to see the Elephant
(Though all of them were blind),
That each by observation
Might satisfy his mind.

The First approach'd the Elephant,
And happening to fall
Against his broad and sturdy side,
At once began to bawl:
"God bless me! but the Elephant
Is very like a wall!"

The Second, feeling of the tusk,
Cried,–"Ho! what have we here
So very round and smooth and sharp?
To me 'tis mighty clear,
This wonder of an Elephant
Is very like a spear!"

The Third approach'd the animal,
And happening to take
The squirming trunk within his hands,
Thus boldly up and spake:
"I see,"–quoth he–"the Elephant
Is very like a snake!"

The Fourth reached out an eager hand,
And felt about the knee:
"What most this wondrous beast is like
Is mighty plain,"–quoth he–
"'Tis clear enough the Elephant
Is very like a tree!"

The Fifth, who chanced to touch the ear,
Said–"E'en the blindest man
Can tell what this resembles most;
Deny the fact who can,
This marvel of an Elephant
Is very like a fan!"

The Sixth no sooner had begun
About the beast to grope,
Then, seizing on the swinging tail
That fell within his scope,
"I see,"–quoth he–"the Elephant
Is very like a rope!"

And so these men of Indostan
Disputed loud and long,
Each in his own opinion
Exceeding stiff and strong,
Though each was partly in the right,
And all were in the wrong!

This article will address one specific library building which was classified with sick building syndrome (SBS) in March 1988–some nineteen years after its construction. Employee complaints of symptoms brought on by the building have been taken seriously, and attempts have been made to control the building's climate in order to alleviate symptoms, yet they continue to this day.

Our experience of dealing with SBS has been comparable to that of the wise men and the elephant in the poem at the top of this article, and it has taught us three things that I've discovered are common in all cases of SBS. First, the condition manifests itself in many different forms, but basically one may conclude that if a large portion of the building's inhabitants complain of problems–either physical or mental–then the building is suspect. Next, each person reacts to the environment in different ways (see Appendix); therefore no blood test, lung test, allergy test, or any other generalized test could reveal a "common thread." This baffles doctors, lawyers, human resources administrators, and building engineers, yet it is the foundation of human nature to react individually to one's environment. Finally, unless the underlying cause of SBS is simple and easily addressed (which often it is not), the answer is to clean the building, and if the problem persists, clean, clean, and clean again.

We were made aware of these three basic concepts early on in our adventure, yet we didn't want to believe them. We wanted a quick fix, an easy cure, a common symptom, a common causation. While we haven't eliminated the SBS symptoms in our building, we have learned over time how to manage an ongoing problem created by the physical nature of our building's HVAC system. Fortunately, many building managers will have an easier solution to the problem, although the ease with which the problem is fixed can depend on issues pertaining to actual human suffering, the seriousness with which management treats the problem, and the complications caused by dealing with outside professional firms.

Before we begin our adventure, let's go back to 1968, when our library building was designed. A theory then existed among architects that the airways for heating and air conditioning a building could be made part of the structure. During construction of the library, as the cement for each floor was poured, a hollow space of 5 to 8 inches in height was created between the ceilings of lower levels and the floors of the levels immediately above them. Thus, with careful addition of metal spacers to denote different zones, large floors could be built with vents spaced evenly in the ceiling to provide climate control. Air from basement HVAC fans could then be blown through a hollow wall (or in a few cases through traditional metal ductwork) and channeled into the hollows. The idea was to create a quiet air conditioning system along with a form of climate control that used the properties of the large masses of concrete to provide a steady temperature in the building.

The building is a massive rectangular structure of four stories. The lower story, which is partially below ground on one side, has conventional metal ductwork. The upper three floors are comprised technically of two buildings connected by a three-story lobby housed under one roof. The larger "building" holds three levels of bookshelves, seating, and meeting rooms. The poured-concrete interstitial spaces are in this section. Across the lobby the other "building," which is set a half story lower, has mostly metal ductwork. This unusual design exists in order to comply with a requirement for a flat roof with no cooling towers or elevator motor rooms, since the roof could be easily seen from a nearby campus tower. To compensate for the prohibition of cooling towers on the roof, a 12′ × 12′ air intake vent was constructed below ground level in the north wall of the building, and the large pit containing this vent was covered with a metal grate.

As a whole, this library building is handsome, and it has proved to be both flexible and useful over the years. Even though it is nearly 40 years old, it gives many visitors the impression that it is a new building. Because of its size and its grand spaces displaying art and artifacts, the cost of demolishing it in order to completely remove the source of SBS would be prohibitive.

Our story begins on March 18, 1988. You may wonder how I can be so precise. The fact is that I had had an operation which kept me away from work and away from the building from July until February. Upon my return, I suggested that the staff keep a daily log diary so that we could show various administrators the complex tasks that we performed every day. So, on March 18, I entered into the log that a majority of the staff (we only had about 12 full-time people) reported that "they could not breathe." I immediately called our physical plant crew. They immediately and dutifully reported back that all the HVAC pumps and fans were working correctly. Yet, the staff continued to complain, and I called for help again. This time, the physical plant crew discovered that the massive air intake for the building was clogged with dead leaves that had fallen through the grate and been caught in the intake fans.

A young HVAC crew member promptly removed all of the debris and then was home for a week with a respiratory illness. This should have been our first clue. In any case, the breathing problem in the building abated somewhat, but complaints continued. Being librarians, the information services staff began to search the literature for solutions. In 1988, little information about sick buildings existed. Nevertheless, the librarians found references to a company, ACVA Atlantic (Air Conditioning & Ventilation Associates Atlantic), of Fairfax, Virginia, that specialized in the treatment of sick buildings. That May, this company conducted a thorough inspection and analysis of our building. In preliminary discussions with them, we learned our first lessons about SBS. First, all people react to their environments differently, and there is no single test that will diagnose SBS. Also, we learned that even if the problem is solved, it may recur, thus beginning the mantra of clean, clean, and clean again. Although ACVA was a knowledgeable and professional team fully familiar with various sick building situations, our library staff faced another problem: skepticism in our community from administrators, physical plant staff, and another department that occupied the section of the library that was ventilated with conventional metal ducts.

Fortunately, the top administrators, though new to the concept of SBS, fully funded and supported all efforts to address the problem, and in fall 1988 the building was shut down and all of the ducts were treated with a biocide. The relief was immediate and we thought that all our problems were over. We also noted that some of the staff reacted more violently to the air than others.

However, over the following weeks, symptoms returned to some of the staff, and in March 1989 all library personnel were attended to by several physicians hired by the school. One doctor found that there were four molds present in common among the staff, but in varying degrees, so one year after first noting our problem we brought back ACVA Atlantic for another inspection of our building. In May and July of 1989 we had all the ductwork treated with a biocide, both times with diminishing short-lived results, and over subsequent years we used a number of other professional companies to clean, clean, and clean again. This is not to indicate that any of the companies were inadequate; it was just that we did not want to see the elephant in the room. The elephant was, of course, the original construction design that required air to be blown through a large ventilation system encased in concrete. The cost of bypassing all of this with a new air conditioning system or the unrealistic cost of tearing down the building and starting all over again was why we persisted in approaching every avenue of remediation via cleaning the HVAC system:

- in July 1989 we brought in Healthy Buildings International for another opinion and inspection
- in September 1989 we built a new outside air intake shaft that opened eight feet above ground and bypassed the below-ground chamber from which our problem originated
- in January 1990 we closed the library, cleaned all of the ductwork with giant powerful HEPA vacuums while continually filtering the interior air, and sprayed a hospital-grade biocide throughout the building.
- in the fall of 1990, we removed the heavy wool wall-to-wall carpeting that covered all of the study and stack floors. One telling story here is that as the carpet was removed from the former special collections display room, you could literally stand in the part of the room where the carpet was removed and breathe more easily than when standing upon those sections still in place. We sensed that the carpet was loaded with dust and spores from 20 years of use.

In October 1990, after not getting satisfactory results from our cleaning efforts, we decided to approach the problem somewhat differently. We brought in Monsen Engineering to do a complete analysis of the HVAC system. They concluded that the chillers currently in use could not possibly cope with the requirements of the structure and that furthermore, the air flow change in the building was woefully inadequate. One year later, heavy-duty air filters were installed in the building's HVAC system and new hardwood and vinyl tile floors were installed throughout the library. All of these projects consumed lots of money and often required serious displacement of staff, students, and equipment in order to be accomplished.

When employees' symptoms persisted, we tried another tack: Bring in the fresh air. Our modern building, to allow the HVAC systems to work efficiently, had no windows that opened. In fact, it had massive, thick, bronzed glass set in Corten steel frames. We began a project of installing in various offices and other places windows that opened. They provided a measure of relief so we added more–again at considerable expense; however, over time we realized that we were making things worse. Fresh air blowing in was unfiltered air, and in the summer we were introducing very moist air that created further problems as it was sucked back into the return air ducts where mold could then grow again.

After our HVAC remedies failed, in February 1995, we brought in Dr. Wallace Rhodes of ENACT, a company in Atlanta. Again, his firm did a complete analysis and study, first doing it blind and then comparing their results with the reports of prior professional firms. His reports concurred with prior reports and everyone involved with this study was in agreement, yet no single solution seemed to emerge. In September 1995, some seven years and six months after our initial outbreak, we held an extensive planning meeting with administrators, professional consultants, engineers, and staff to determine how we should proceed.

In January 1996 we closed the entire library again. We created a satellite library in another campus building with staff, computers, and copiers, but no books. We used runners to obtain books from the closed facility. In the library, special robots crawled the ductwork, removing all kinds of things, including original construction debris. These robots then sprayed and sealed all the concrete and metal duct surfaces with a biocide-treated paint. This accomplished two things: it helped seal the porous concrete, and video cameras on the robots showed us that all surfaces, to the best of the contractor's ability, had been cleaned and treated. This massive task was followed by engaging ISS Energy Services to completely clean the interior of the building–every book, every surface–with HEPA vacuums and special cleaning materials. The building air was retested, and it was found that the various contaminants had dropped by factors of 100, 1,000, and 10,000. This was good news.

In August 1996 all the unusual metal discs that allowed air through holes in the concrete ceiling were rebalanced by a professional firm, and in September and the following April, two additional analyses of the air quality and status of the fans, ducts, filters, and systems were conducted. These examinations led school officials to the conclusions that (1) the building's chillers were inadequate for the library and needed to be replaced with new units of proper capacity, which was done; (2) the building tended to run on negative air pressure, resulting in infiltration of the structure (this was temporarily addressed by throttling the exhaust air dampers); (3) the maintenance staff needed to be vigilant in cleaning both the HVAC hardware and the interior work spaces, which was initially done but which has gradually lessened as different cleaning contractors have been hired and the knowledge of procedures has been lost over time; and most importantly, (4) the building should have new ductwork to bypass the HVAC design that was creating the problem.

As fate would have it, though, one of the factors affecting the replacement of the original system with new ductwork was the fact that during construction an asbestos-containing product called Amosite was used to fireproof the ceilings of the mechanical rooms and other spaces. Before any heavy construction could take place within the building, the Amosite would have to be removed at great expense. Nevertheless, in 1999 it was decided to once again close the building and remove all of the asbestos on the mechanical room level, including all of the offices and computer rooms. At the same time, a new air-handling system was installed for this entire lower floor. And, for good measure, all offices were repainted with antimicrobial paint. The university HVAC engineer designed a system that would manage the problems (at least for one level) and prepare the way for his eventually installing new ducts in the upper areas.

Moreover, wherever possible, separate areas within the building were given their own air-handling systems and separate air intake ports. This was not always without its own problems. In 2001, an outside firm was engaged by one of our tenants to put in an entirely new independent system fed from the roof into a large area of new offices occupying space that had formerly held the library's journal collection. The firm overestimated the cooling capacity because they failed to consider that the ceiling of their new offices was already cooled by air flowing through the hollow interstitial space designed for the original stack area. The result, once summer arrived, was condensation that allowed mold to grow abundantly on walls and ceiling. This is an unfortunate example of engineers not truly understanding their environment, and the problem was not corrected until our own building engineer was able to duct some of the excess cooled air into other stack spaces.

In 2003, we brought in yet another firm, Atlantic Environmental Inc., to look at how things were going in the building. Although we still regularly had staff complaints, things were generally much better. Having reviewed the past studies, this firm did its own series of repeat tests. They also thought that with all of the hidden concrete duct surfaces there might be lime or calcium dust as a factor. Those tests proved to be negative. Therefore, their report once again suggested that the proper cure was to eliminate the use of interior walls and floors as a means of transporting the air to the working spaces. New ductwork bypassing everything is necessary. And, when new ductwork is installed then new HVAC handling equipment with modern filters, humidification, and dehumidification equipment should be installed too. To date, this work has not been done due to the capital expense and other priorities.

Now let us review what has happened after allergists, doctors, multiple firms specializing in sick buildings, cleaning firms, changes in the provision of heating and cooling systems in local areas within the building, and various other solutions have been attempted. A great deal of money has been spent to date, and it would have been nice to have had a proper cut-and dried diagnosis very early on. This diagnosis would have concluded that the only solution was to stop using the concrete ducts that were integral to the building structure and deliver the air through traditional metal ductwork.

Going back, it should be noted that the building, because of its design, tends to switch between extremes of hot and cold. It has a three day lag time because of the mass of concrete that needs to be heated or cooled; therefore, if, for example there were a sudden warming trend in the middle of February, it might take at least three days to lower the temperature in the building to a comfortable level. Staff complaints about being uncomfortable often confused the issue of the sick building. Maintenance people were quick to grab on to the idea of the easy fix, and if the complaints indicated, for example, only that employees were too cold, the physical plant workers would turn up the temperature and consider the problem solved. Yet, the symptoms displayed by various personnel were, and continue to be for some, both serious and debilitating. Since the outset of the problem, I have not been able to work in the building on a full time basis. Fortunately, my campus residence is only a few hundred feet from the library. Computer networks, fax, and telephone make it possible to be "in the office" without being in the building.

Other staff have been moved out, moved back, and moved to special rooms in efforts to cope with their symptoms.

To add to our frustration, and despite documented evidence from a number of staffers over the years, we would still hear, "It is all in their head" voiced by people who did not understand our situation. For the doubters, we had many stories from salespeople and day visitors to the building. One of my favorites concerns a large work room that had been cleared of all copy machines, laser printers, and other devices that might be contributing to the problem. Special giant HEPA filter units were installed and run at full blast each evening. Special HVAC systems were installed to bring in fresh air and moderate the temperature. An older retired businesswoman who worked for us part time was sitting at her desk in the morning. She reported that another staff person walked into the room, immediately grabbed her throat, and said, "Ooh, I feel like I am going to throw up." Shortly after this, another person walked into the room and said, in effect, the same thing. This older observer said nothing, assuming that perhaps this was a fluke or that they had somehow triggered each other. However, the day went on and at about 3 p.m., the deputy director, who had been out of the building all day, walked in for the first time, and she said, "Ooh, I feel like I am going to throw up."

Also, we had observed that on humid days, the building tended to be worse. If the lawns were sprayed with weed killer, the building was worse. If there was a sudden deep chill in the outside air temperature, the building was worse. Over time, we concluded that interior mold spores might have been killed off in those situations, and these floating spores triggered reactions in individuals. We have, of course, tried various room and industrial air cleaners, including ultraviolet light devices and other high tech equipment. We also noted, when doing a smoke test, that the HVAC system distributed the air via the ceiling in an odd way. The smoke seemed to drift and hang high at the ceiling like looming clouds in a Steven Spielberg movie. Then it completely filled the space from ceiling to floor and stayed there until the exhaust fans were turned on full blast. In light of these problems, one could easily predict the effect that SBS has had on staff productivity. On days when the staff complains about reactions, our policy is to send them home, allow them extended lunch hours, or do whatever is necessary for them to feel better. Once, an assistant decided to leave the building because she was feeling unwell. She drove to a nearby shopping mall, then discovered that she was so disoriented that she had to call her husband to pick her up. That same morning, a school administrator who worked outside the building stopped in my office. He did not know of our problems; however, he reported later that after he left my office and started walking back to his building, he grew dizzy and disoriented. These symptoms, as they did for all the staff, disappeared from these two once they spent a few hours away from the building. My own wife always knew when I was in the building for too long, because when I would come home, I would have to sit on the couch, almost propped up, for about two hours before my voice would come back and before I had any energy to do things.

How do we manage SBS now after almost twenty years? We watch constantly for overt signs of staff complaints and reactions. In minor cases, we send the staff home. Sometimes, with networking, they can even work from home. In all cases, we alert building maintenance services to modify current HVAC settings, and check the climate in the building. Documentation is also essential. We have records going back to the time of our initial problem, including daily report sheets filed by each staff person who has suffered symptoms in the building. We keep all key people, including the human resources department, in the loop with regard to major situations. And, always, we hope that someday soon about three million dollars can be found to eliminate the fundamental cause of our problem. To repeat, this fundamental cause is a building so constructed that a major portion of its air conditioning is delivered through concrete vertical ducts and horizontal sandwiches of floors and ceilings with an air space in the middle.

doi:10.1300/J118v25n03_10

APPENDIX

No one symptom can be found in various individuals. Here is a list of some of the reported problems (both past and in some cases current).

Person 1

- voice change
- great deal of mucus in the throat
- coughing (must take antihistamines at night to sleep)
- sharp pains at top of chest, especially when inhaling
- feeling of nausea at bottom of throat
- dizziness, disorientation, extreme irritability, and irrational reactions
- weakness in legs, extreme unaccountable exhaustion,
- loss of concentration and inability to process even simple work tasks
- dryness of lips and mouth, metallic taste in mouth
- tiny poison ivy-like pimples on skin, especially hands
- symptoms clear after time away from building, sometimes in hours, sometimes days

Person 2

- ringing in ears, deafness
- numbness around mouth
- weakness in legs
- congested head, nose, chest
- sore throat, rough deepening voice
- pain in upper chest
- cough
- depressed
- extreme temper
- problem concentrating
- extreme fatigue
- severe athlete's foot on one foot and scalp eczema that would clear on vacation but come back upon return to the building
- dizziness
- problems with coordination
- shortness of breath
- heart palpitations
- pain in back of shoulder and down the right arm
- nausea from the throat and chest, not from stomach
- pain in abdomen that would disappear after leaving the building

Person 3

- headaches, nausea
- heart palpitations
- depression
- rash diagnosed as eczema
- mood swings
- pressure in both ears
- loss of balance, especially on first floor
- stuffy nose

- tightness in throat
- rash on hands and foot
- constant nagging soreness down left side of neck, sometimes traveling down arm
- shortness of breath
- pain in chest
- disoriented when writing
- queasy feeling in stomach

Person 4

- nausea
- eyes burning
- ears ringing and clogged
- dizziness
- joints aching, stiff neck
- exhaustion
- very dry, sore throat
- skin on hands peeling away in thin layers

Person 5

- chest and arm pain
- sore throat
- neck pain
- coughing, dizziness
- lightheadedness
- mood swings
- disoriented
- metallic taste in mouth
- dry nose
- eyes feel heavy
- symptoms clear when away for 24 hours

Person 6

- headaches
- dizziness
- coughing
- slight but increasing pain in chest
- numbness in arms
- extreme dryness of throat at times
- total drain of energy which occurs as early as one hour being at work
- nausea
- disorientation and confusion
- rash on foot that would get very itchy when in building

Person 7

- very dry mouth
- queasiness
- eyelids felt itchy and one eyelid was sore and swollen

Major and Minor Mold Outbreaks: Summer of 2002

Nancy E. Kraft

SUMMARY. In the summer of 2002, the University of Iowa Libraries Preservation Department responded to an unusual number of mold incidents, large and small. This article discusses the strategies used and outlines steps to a successful mold remediation. doi:10.1300/J118v25n03_11 *[Article copies available for a fee from The Haworth Document Delivery Service: 1-800-HAWORTH. E-mail address: <docdelivery@haworthpress.com> Website: <http://www.HaworthPress.com> © 2006 by The Haworth Press, Inc. All rights reserved.]*

KEYWORDS. Mold, remediation, University of Iowa Libraries

INTRODUCTION

Mold, a vital component of the natural world, is an unwelcome visitor in libraries. Mold can destroy library collections and make them hazardous to readers and staff. Maintaining an environment with a stable temperature and relative humidity and good air circulation significantly

---

Nancy E. Kraft, MLIS, is Preservation Librarian and Head of the Preservation Department, University of Iowa Libraries, 100 Main, Iowa City, IA 52242 (E-mail: nancy-e-kraft@uiowa.edu).

[Haworth co-indexing entry note]: "Major and Minor Mold Outbreaks: Summer of 2002." Kraft, Nancy E. Co-published simultaneously in *Public Library Quarterly* (The Haworth Information Press, an imprint of The Haworth Press, Inc.) Vol. 25, No. 3/4, 2006, pp. 127-141; and: *Dealing with Natural Disasters in Libraries* (ed: William Miller and Rita M. Pellen) The Haworth Information Press, an imprint of The Haworth Press, Inc., 2006, pp. 127-141. Single or multiple copies of this article are available for a fee from The Haworth Document Delivery Service [1-800-HAWORTH, 9:00 a.m. - 5:00 p.m. (EST). E-mail address: docdelivery@haworthpress.com].

Available online at http://plq.haworthpress.com

doi:10.1300/J118v25n03_11

decreases the chance of a mold outbreak. Monitoring the collections and environment, establishing standard protocols for processing collections, and following current mold remediation practices will provide a safer environment for staff and readers, reducing the number of mold incidents and exposure time to mold.

In late summer of 2002, the University of Iowa Libraries Preservation Department responded to an unusual number of mold incidents, large and small. The following narrative outlines the steps needed during a response to a mold outbreak.

MAJOR OUTBREAK: UNIVERSITY LAW LIBRARY 5,000 VOLUME RARE BOOK ROOM

On Friday the University of Iowa Law Library staff noticed the relative humidity in the Rare Book Room was at 80%, set up dehumidifiers to bring the humidity down, and went home for the weekend. On Tuesday staff noticed mold while retrieving a book from the room and immediately contacted the Preservation Department, asking for assistance.

Having already responded to several false "mold alarms," I assumed that this was another one and did not ask any questions. The assistant conservator and I went to assess the situation with no protective gear and walked into a room with mold on the carpet, walls, chairs, and books (Illustration 1). Even our brief exposure without protective gear was hazardous. The assistant conservator developed a sore throat and brief fever and I was ill for six weeks. However, the law librarian was in and out of the room several times without any protection yet without any adverse side effects.

We returned with protective gear and our digital camera for documentation, and assessed the situation. During our assessment, we checked to see if there were any windows that could be opened (there weren't), determined where the vents were, looked to see if any aisles were blocked or difficult to access, verified that there was mold growth throughout the entire area, noted the current temperature and relative humidity, and took several photographs of areas with mold. Our assessment gave us enough information to begin discussions with key players to determine strategy and the digital photographs gave visual confirmation to the extent of the mold growth.

As with any disaster response–and mold covering rare books was certainly a disaster–it was critical to take time to determine a strategy

ILLUSTRATION 1. Books in Law Library Rare Book Room with Mold

Image ©University of Iowa Libraries. Reprinted with permission.

for stabilization and recovery. The damage was done; there was mold everywhere. A little more mold growth wouldn't do as much damage as a poorly conceived plan. Taking time to assess and document the situation, do a little research, and identify and consult key players would actually shorten recovery time.

In addition, to have a successful recovery it was critical to understand the building air circulation system, internal politics, and bureaucratic procedures. When discussing the situation and possible solutions we listened for tip-offs for possible roadblocks. I learned that the law library area facilities person was slow to respond and the building mechanic felt that his opinions and input were not valued. I consulted with our area facilities person, who assisted with by-passing all the bureaucratic reporting levels for a fast response without creating additional problems. I also made sure that the building mechanic was included in all relevant discussions.

As the preservation librarian, I wanted to be a part of the recovery process and not sidelined so I offered to coordinate remediation effort. Since I had worked with the remediation company before, I volunteered to call them. These two actions allowed me to play a key role throughout the remediation process. As the coordinator I pulled key people together for meetings, kept meetings to a minimum, and kept everyone informed via email.

By late Tuesday afternoon a university engineering team was working with the building mechanic to stabilize the environment, the remediation company had assessed the situation, set up a large dehumidifier and consulted with the company's senior scientist, the law librarian had consulted with the finance manager, and I had arranged for key people to meet Wednesday morning to discuss the situation and determine action steps.

The mold remediation team consisted of:

- Library director and assistant–made decisions, gave permission to action items directly impacting rare books, use of library staff, and after-hours activities.
- Business manager–approved or got approval for actions requiring funds.
- HVAC mechanic–provided information on how the heating, ventilation, and air conditioning (HVAC) system worked.
- Mold remediation manager and supervisor–had training and appropriate equipment, skills, and staff to handle actual remediation work.
- Libraries facilities manager–advised preservation librarian on the ins and outs of working with facilities engineers.
- Assistant conservator–provided digital photographic documentation, set up environmental monitors, researched mold growth on books.
- Conservator–provided training on appropriate cleaning and handling of books.
- Preservation Librarian–researched latest mold remediation practices, coordinated and monitored remediation.
- Risk manager–reviewed remediation contract with vendor, advised on legal and safety issues.

The only key position not represented in the above team was a public relations person. Since we were between school sessions, we did not anticipate any media interest and decided that any media inquiries would be directed to the library director. We did not get any inquiries.

We held our first formal meeting Wednesday morning. Most of our discussion centered on how to set up mold remediation without further contaminating the building. In an ideal situation we would start venting the air from the Rare Book Room to the exterior. This was not an option as we could not find a direct route to the outside. After much debate, we decided to add extra filters to the HVAC system and adjust the positive/negative airflow to vent air out of the room. We felt that this procedure would provide a safe work environment in the rest of the building. However we did give staff permission to stay home if they were concerned about any health risk.

We set action items, timeframe, and responsibilities:

- Mold remediation manager would consult with the company's senior scientist.
- Preservation librarian would request exemption from energy curtailment. In order to save money, the university had an agreement with the power company to partially shut down or curtail its cooling system when requested.
- Facilities staff would request air sampling. It was common practice to take an air sampling from the contaminated room or building and compare it to an air sampling from the outside environment. Once remediation was over another sampling would be taken to demonstrate that the inside environment had been stabilized or at least was as good as the outside air. S*ince Hurricane Katrina this practice is being re-thought. Air sampling comparisons from inside and outside proved impractical in the aftermath of Hurricane Katrina as the outside air was just as contaminated as the inside air. Air sampling is only needed if there is uncertainty as to extent of contamination throughout the building. Generally a visual inspection is sufficient.*
- Facilities would request identification of type of mold. (The remediation company's protocol required that the type of mold be identified. We could do it through our lab or through theirs.) Identification of mold type was fairly standard practice but has been discontinued. *The strategy for mold remediation is the same regardless of mold type. Establishing the type of mold is expensive and remediation needs to start before the type is known. It's less expensive, and faster turn around time, just to ask for verification that the outbreak is, indeed, mold.* For the record the mold type was penicillium.
- Remediation supervisor and crew would meet with the conservator for training on cleaning books. Even top notch remediation

companies tended to hire temporary help and liked to use biocides. We wanted to insure appropriate handling and cleaning of the books and insure that no biocides or any type of mold suppressant products were used on the books.

- Law librarian would keep staff informed.
- Preservation librarian would coordinate and keep everyone informed via email.

Mold remediation began 4:00 p.m. on Wednesday, slightly 24 hours after the mold outbreak was reported to the preservation department. The remediation company was hired with a contract approved and signed, the room was sealed off, additional filters were in place, air pressure changed to vent air out of the room through the HVAC duct work to the outside, air samplings taken, scientist consulted, expenditures approved, and a large dehumidifier in place and operating. All these actions could not have been accomplished and in place without a team committed to working together to get everything in place as quickly as possible. We were lucky that everyone understood that stabilizing the situation and beginning mold remediation as soon as possible were important and everyone made it his/her top priority; everything else went on hold.

Midmorning Thursday, the remediation manager reported that the RH (relative humidity) was actually increasing. Internally the books were still dry so we weren't at a crisis point but corrective action was needed. The building mechanic adjusted the positive air pressure and the company brought in a bigger dehumidifier. I looked for storage for the books just in case these actions didn't start bringing the humidity down. By noon the area had stabilized and the humidity began to drop.

While our vendor worked to stabilize and begin remediation, we met to try to figure out the source of the problem. Mold remediation is time consuming and expensive; it was an experience we preferred not to repeat. During our discussions we learned that the HVAC system was being run improperly; the reheat was turned off due to a directive several years ago in an effort to conserve energy and save money. This directive had since been rescinded but had never reached the building mechanic. The reheat dries the chilled water before circulation into the interior. Running the system incorrectly had always frustrated him; he was more than willing to work with the university engineers to correct the HVAC system problem.

The system had been running incorrectly for years without any mold outbreaks so why an outbreak now? Although we'll never know for sure, we suspected it was a combination of two events. We had several days of exceptionally high outside humidity which increased the inside humidity since the library had no system in place to control the humidity. School was out of session making it a good time to clean carpets–including those immediately outside the Rare Book Room–adding even more humidity into the air.

By Friday morning or a day and half after we began remediation, the RH was at 42% from a high of 76%. We now had an enclosed environment that was no longer conducive to mold growth. I requested that the remediation company report temperature, and relative and absolute humidity readings to me at regular intervals. These readings gave me a picture of what was happening environmentally and kept me informed as to the progress of stabilization.

• Wed. p.m.	78 °F	60% RH	89 grains
• Thurs. a.m.	78 °F	73% RH	106 grains
• Thurs. p.m.	82 °F	45% RH	73 grains
• Fri. a.m.	79 °F	42% RH	62 grains

The piece of information that I was unfamiliar with was the "grains," the absolute or specific humidity, given in grains of water per pound of air. This information was used to determine dew point or "danger zone" when water would begin to condensate–something to be avoided. A psychrometric chart was used (Illustration 2) to determine dew point and drying strategy.

In the meantime we were having trouble getting the HVAC system to function properly. Someone finally figured out that the problem was due to work construction outside the building that had inadvertently shut down part of the system and not the equipment itself. Our only option was to run the system too cold. We sent out an alert to staff to bring their sweaters to work. Unfortunately, this wasn't our only setback. On Monday, we learned that clean up on the rare book room had been put on hold because the auditorium in the same building had mold which needed to be remediated in time for the fall session. Since we had the situation stabilized and the room sealed, it made sense to work on the area that was of higher use and had more immediate need.

Our conservator worked with the remediation company to establish cleaning procedures. Since the books were rare and some quite fragile, he requested that the books be cleaned without removing them from the shelves. This proved to be impractical because the mold had gotten in between the books and onto the shelves. All the books had to be taken off the shelves and cleaned. Training on proper cleaning and handling

ILLUSTRATION 2. Section of Psychrometric Chart

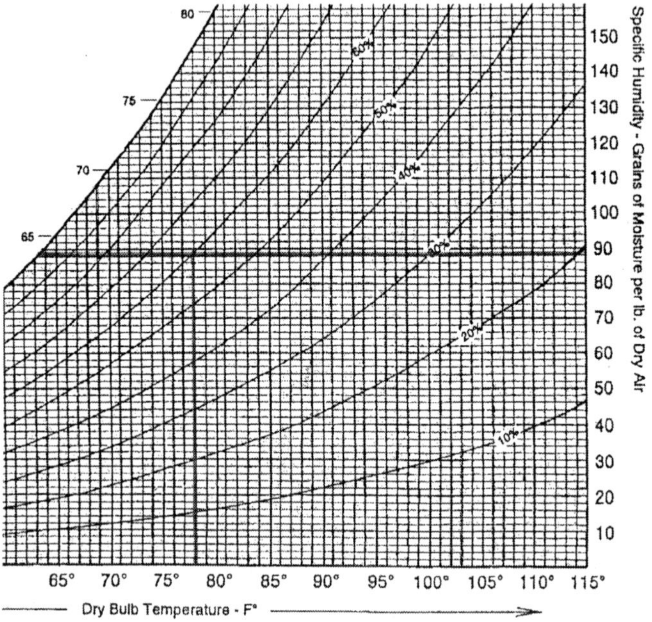

of books was provided by the conservator. The books were vacuumed and then wiped down with a dry sponge. Fortunately a visual inspection by conservator determined that mold had not gotten into the interior of any book, so only the outside exposed edges and covers needed cleaning. Books that needed special attention were flagged for the conservator.

The shelving and walls were wet sponged with phenolic biocide. The biocide treatment was not necessary. Mold can not be completely removed even with a biocide treatment. When it dries up or if wiped off, it leaves an external amorphous film which adheres to the substrate so strongly that it can not be removed (Illustration 3). Using the biocide was standard company protocol; we decided that it was safe to use the biocide on the shelving and walls. We did not allow them to use biocide on the books as biocide solutions could damage the books. *The use of biocides is no longer recommended for most mold remediation. Vacuuming and dry sponge cleaning is acceptable for cleaning books and documents. Vacuuming and/or water and detergent are recommended*

ILLUSTRATION 3. Mold Biofilm

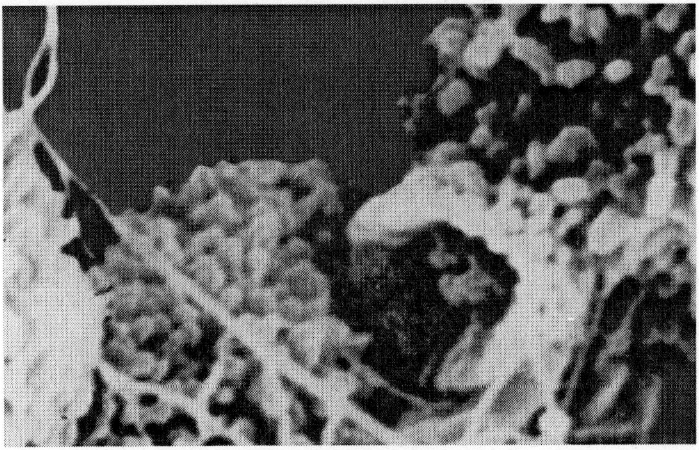

Provided by Mary-Lou E. Florian. Reprinted with permission.

for most everything else. Using anything else introduces potential health problems and possible structural changes and damage to the materials being treated.

Once everything was cleaned, we checked to make sure there was no mold growth in the ductwork. The ductwork was given a clean bill of health. We were through with remediation and could go back to work. Well, not really. We established a monitoring system. Six weeks later, the conservator discovered about 30, very scattered, old buckram covered books with mold. The conditions at the time were 72 °F and 36% RH. There have been no mold outbreaks since then.

The total cost for the mold remediation was $50,000.

MINOR OUTBREAKS: UNIVERSITY MAIN LIBRARY

We had several small outbreaks during the summer of 2002. In the Special Collections area, mold appeared on about 200 portfolio enclosures (Illustration 4) that were probably made in the 1970s. They were all constructed from the same type of cloth and probably used the same adhesive. A few items that were shelved along side the enclosures also had some mold. We're assuming that something in the materials used to construct the enclosures made them more susceptible to mold growth.

ILLUSTRATION 4. Mold on Portfolio Enclosures

Image ©University of Iowa Libraries. Reprinted with permission.

The source of the problem turned out to be a vent that was not working correctly. We decided to keep the enclosures rather than replace them. We thought they could act as a "canary"–a warning that we had an environmental problem. Better to have mold on book containers rather than on the books. We put the enclosures in an isolated room, cleaned the enclosures with a HEPA (high efficiency particulate air) vacuum cleaner and lightly wiped the enclosures with alcohol, reboxed the books and returned them to Special Collections. We did confirm that the growth on the books was mold, identified as sepedonium.

Our "canary" theory did not hold up. In the summer of 2006, we had another small mold outbreak in the Special Collections area. This time the mold was on books with a particular type of binding. The event was triggered by a malfunctioning HVAC which was blowing humidity into the area. After the system was stabilized, we cleaned the books. However, mold outbreaks continued, scattered throughout the area, on a few books here and there. There did not seem to be any discernable pattern or particular area. The humidity was up but we couldn't figure out why. Suspecting that the problem might once again be a vent, we mapped the area, detailing vent locations and mold growth areas (Illustration 5). We discovered that more than one vent was not working correctly. We worked closely with our facilities staff. They supplied us with an accurate floor plan of the area, we mapped the outbreaks, and they fixed the vent problems.

Right on the heels of the outbreak in Special Collections, staff discovered mold on books in the university archives (Illustration 6). This time the mold was aspergillus. By then we were feeling slightly crazed dealing with all the mold incidents. One afternoon, we had our lab coats and face masks ready to check out the mold situation and started singing, "We're the moldbusters!" There was no obvious reason for the outbreak. The temperature and relative humidity had not increased and all the HVAC equipment was functioning correctly. The area where the outbreak occurred was the book range at the end of the venting system in the area. Even though the fans were working properly, we speculated that that particular area was not getting sufficient air circulation. Our building manager had the vents cleaned and there were no more mold outbreaks.

ILLUSTRATION 5. Mapping of Mold Outbreak in Special Collections

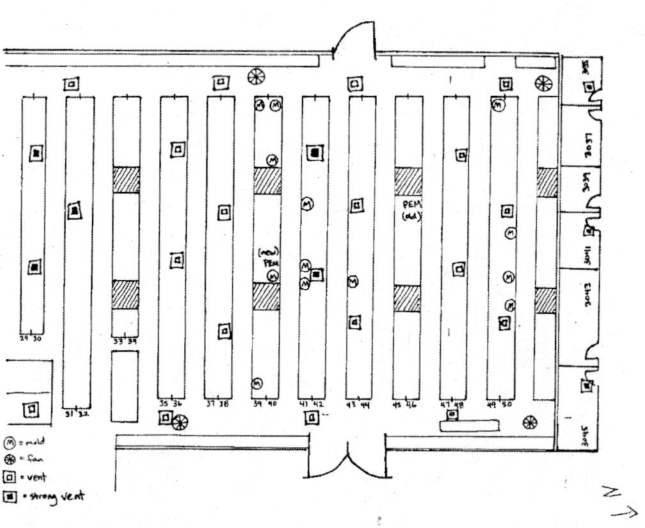

Image ©University of Iowa Libraries. Reprinted with permission.

ILLUSTRATION 6. Books in University Archives with Mold

Image ©University of Iowa Libraries. Reprinted with permission.

We're not sure if the vent cleaning corrected the problem or seasonal lower outdoor humidity did. As an extra precaution, the archives staff changed the shelving arrangement, replacing the books with boxed material of lesser value and shelving the books in a different area. In the event of another mold incident in that area, replacing boxes would be much easier and simpler than cleaning books.

One of the preservation staff spotted a shelf of damp, moldy books in the cataloging area. They were gift books that had arrived in that condition. The bibliographer accepting the gift had set them out to dry in a heavily staffed area. We explained to him that mold could be a health hazard and any items suspected of having mold should always be brought to the conservation lab for treatment. We sent out a staff advisory to contact the preservation department whenever material had evidence of mold. We also met with acquisition staff to discuss the need for staff protection when processing dirty, dusty collections.

We capped our season of mold with the "smelly books" incident. During a heavy rain downpour, on a Thursday evening some of the basements flooded in Iowa City. Unfortunately for us, one of the flooded basements had 75 university library books. The student contacted the library the next day for advice on how to handle the wet books. He was advised to bring the books in immediately and even though it was close to closing time, staff would wait for him. Although staff waited long after 5:00 p.m., he didn't bring the books in until Monday. By that time the books not only had the debris–bugs, leaves, twigs, mud–left by the floodwaters but also mold and were quite smelly.

Our library had no space available that could be designated as a containment or intake area. The smell was too strong to leave in any regular work area. In fact the smell was too strong for anyone to work with them. We found the perfect solution–a room on the top floor located next to the air handlers. This room also had a door that could be opened to let in outside air. Due to the proximity to the air handlers, there was strong, constant air circulation–not windy, just really, really good air movement. The room was large enough to set up long tables and spread out all the books for quick drying. Miraculously, the smell almost completely dissipated overnight. Since salvaging the books would be very labor intensive, we worked with the bibliographers to identify any books that could be discarded. It took several weeks to dry, clean, and repair all the books.

IN CONCLUSION

We learned during these intense two months of mold recovery that no two incidents are alike; each requires a different remediation strategy. However the steps to managing a mold remediation are the same:

- Assessment and planning are critical to a successful recovery.
- Identify and contact key players needed for remediation and keep them informed.
- Maintain a close working relationship with the building and facilities staff.
- Verify that the growth is mold but there is no need to identify species.
- Protect yourself during assessment; assume that it is mold until proven otherwise. Illnesses due to exposure to mold can result from a single exposure to a high level of mold, multiple exposures of low level, or even a single exposure of low level mold contamination.

    - Wear an N95 disposable respirator.
    - Use disposable gloves if handling the material.
    - Goggles or protective eyewear should be worn.
    - Don't touch your eyes or mouth if you've touched a moldy item.
    - Wash your hands as soon as possible after vacating infected area.
    - Take a shower and wash your clothes in hot water and bleach.

- Document areas of mold contamination photographically.
- Keep a log or daily journal, documenting the incident.
- Monitor the environment.
- Keep your risk manager or insurance agent informed.
- Maintain good communication with building occupants and all key players.
- Identify why the outbreak occurred and work to correct situation.
- Be prepared for setbacks.
- Solutions may need to be creative and may need to be modified.

Remediation is expensive whether the incident is large or small. Monitor your environment. Be familiar with what the daily readings are so that you recognize a change in humidity and temperature. If the budget doesn't allow for a digital monitor, purchase an inexpensive thermo-hygrometer. There are several types available between $30 and $100. If there is a spike in the humidity, increase the air circulation immediately. *Increasing air circulation is the single best thing you can do to reduce a chance of a mold outbreak.* To grow, mold needs to stick to a surface. A good airflow will reduce its chances of sticking.

The typical household model dehumidifier will not assist in reducing the humidity because it is designed to reduce humidity to 60% and not any lower and the coils will freeze and stop working once the room temperature falls below 68°. The correct type of dehumidifier is expensive; money is better directed to the purchase of good, heavy duty fans. If a decision is made to purchase or rent a dehumidifier, find out the rate of water removal in pints per hour, the temperature it can handle before coils freeze and to what level it can dehumidify.

Maintain an up-to-date list of key players so you are prepared in the event of a mold outbreak. Smaller organizations may need to rely on board members or members from within the community or call on individuals from a state university, state historical society or similar entity. A regional conservation or preservation center is also a good resource.

BIBLIOGRAPHY

Damico, Andrew. "My Battles with Mold." *Archival Products News*, 12.2, 5-8. <http://www.archival.com/newsletters/apnewsvol12no2.pdf>.

Patkus, Beth Lindblom. "Emergency Salvage of Moldy Books and Paper. Technical Leaflet." Northeast Document Conservation Center. <http://www.nedcc.org/resources/leaflets/3Emergency_Management/08SalvageMoldyBooks.php>.

Florian, Mary-Lou E. *Fungal Facts: Solving Fungal Problems in Heritage Collections*. London: Archetype Publications, 2002, reprinted. 2004.

"Guidelines on Assessment and Remediation of Fungi in Indoor Environments." New York City Department of Health and Mental Hygiene. 2006. <http://www.nyc.gov/html/doh/html/epi/moldrpt1.shtml>.

"Indoor Air: Mold." U.S. Environmental Protection Agency. 2006. <http://www.epa.gov/mold/index.html>.

Kraft, Nancy E. "Moldbusters Thrive in the UI Libraries Preservation Department." *Bindings* (Fall 2004): 4. <http://www.lib.uiowa.edu/about/bindings/fall2004.pdf>.

Morris, Pat. "Coping With a Mold Outbreak in the Archives." *Archival Products News*. 10.3. 5-8. <http://www.archival.com/newsletters/apnewsvol10no3.pdf>.

Southwell, Kristina, "Chlorine Dioxide: A Treatment for Mold in Libraries." *Archival Products News*. 10.3. 1-4 <http://www.archival.com/newsletters/apnewsvol10no3.pdf>.

University of Minnesota. Environmental Health and Safety. 2001. <http://www.dehs.umn.edu/iaq/fungus/>.

Additional resources may be found at: CoOL: Conservation OnLine <http://palimpsest.stanford.edu>.

Psychrometric chart and calculator may be found at: <http://www.p1m.com/PSY-Chart.pdf> and <http://www.uigi.com/WebPsycH.html>.

doi:10.1300/J118v25n03_11

Keeping the Beast at Bay: Fighting Mold at the University of Missouri-Columbia Journalism Library

Cynthia Dudenhöffer

SUMMARY. This article illustrates the course of action taken by the staff of the University of Missouri Journalism library during a mold outbreak in the Fall Semester 2006. While the parent library system, the University of Missouri-Columbia Libraries had a disaster plan in place, multiple issues occurred that subsequently changed the course of action over the 3 weeks of the cleanup. Those involved took swift action to save a unique journalism collection, and the ultimate results were positive. Consequences of these decisions are discussed. Suggestions for other libraries facing similar circumstances are provided. doi:10.1300/J118v25n03_12 *[Article copies available for a fee from The Haworth Document Delivery Service: 1-800-HAWORTH. E-mail address: <docdelivery@haworthpress.com> Website: <http://www.HaworthPress.com> © 2006 by The Haworth Press, Inc. All rights reserved.]*

Cynthia Dudenhöffer, MALS, is Director of Information Resources, Central Methodist University, Fayette, MO 65248 (E-mail: cmdudenh@centralmethodist.edu).

During the mold outbreak, she was an Information Specialist serving the Missouri School of Journalism and the University of Missouri Libraries, Columbia, MO.

[Haworth co-indexing entry note]: "Keeping the Beast at Bay: Fighting Mold at the University of Missouri-Columbia Journalism Library." Dudenhöffer, Cynthia. Co-published simultaneously in *Public Library Quarterly* (The Haworth Information Press, an imprint of The Haworth Press, Inc.) Vol. 25, No. 3/4, 2006, pp. 143-150; and: *Dealing with Natural Disasters in Libraries* (ed: William Miller and Rita M. Pellen) The Haworth Information Press, an imprint of The Haworth Press, Inc., 2006, pp. 143-150. Single or multiple copies of this article are available for a fee from The Haworth Document Delivery Service [1-800-HAWORTH, 9:00 a.m. - 5:00 p.m. (EST). E-mail address: docdelivery@haworthpress.com].

Available online at http://plq.haworthpress.com

doi:10.1300/J118v25n03_12

KEYWORDS. Libraries, library disasters, mold, environmental controls, planning, humidity, cleaning

INTRODUCTION

On Wednesday, August 23 2006, during the training of a student library assistant, visible mold growth was discovered in the stacks of the University of Missouri-Columbia (MU) Journalism Library. At first glance, the mold growth seemed to be isolated in the back corners of the stack space, where ventilation and temperature control were practically non-existent, but as library staff examined the stack area more closely, it was clear the damage went far beyond the first estimates. The interim head of the Journalism Library immediately contacted the library director and the journalism building coordinator, and began searching for measures to deal with mold contamination. Throughout the next 3 days, the groundwork was laid, procedures were put into place, and clean-up began. Over three weeks later, the library stacks re-opened to users and all visible mold growth was gone.

BACKGROUND

The MU Journalism Library is a small branch of the MU Libraries system. The main purpose of the branch is to provide information resources and support for the Missouri School of Journalism, a college within the University of Missouri-Columbia. In May of 2005, the Journalism Library moved into temporary quarters anticipating the building of a new library space within the Reynolds Journalism Institute, to be completed in summer of 2007. The temporary space was located in the basement of a building called Neff Annex, which also housed the printing presses of the newspaper produced by the school, the *Missourian*. The rooms converted to hold the library consisted of two converted classrooms for a circulation/reference area and reading room, and the old darkroom, used to house the print collection. Only minimal renovations were done to convert the space and subsequently, the entire area, especially the collection space, was poorly suited for library habitation. The bare concrete floor and walls were not sealed; large holes were left in the ceiling from asbestos abatement. There was no temperature, air movement or humidity control; only one small dehumidifier was located in the space, due to a lack of power outlets. The space was damp, drafty, and confined, and was not highly utilized by the library's potential users, despite the presence of study areas. Still, the rationale was that the library would only occupy the space for about two years, subsequently moving into a state-of-the-art facility.

The specialized collection at the time of the mold numbered around 30,000 volumes, and included a photography collection (with many volumes printed on clay-based paper), an original complete run of *LIFE* magazine, and an historic typography section, to name a few highlights. In order to fit into the space, books were shelved on interior and outside walls, on bottom and top shelves of ranges, and much more tightly than in the previous library space, though this practice is contrary to library policy. About 12,000 items were housed off-site, at the University of Missouri Library Depository. The Journalism branch functioned somewhat independently, but all technical services were performed at the main library.

The summer of 2006 was extremely hot. Record heat indexes upward of 100 degrees were recorded though the first week of August. Neff Annex was cooled by a chilled water system, consisting of cycling cold water though pipes, in order to lower temperatures throughout the space. While this system works, clearly it's not the most effective way to maintain temperature control within a library. In the middle of August, pipes in the first floor cooling system became blocked, and temperatures went up to the low 80s. The collection space, housed underground in unsealed concrete several degrees cooler than the floor above, became extremely damp, with humidity ranging about 70 percent. All these conditions created the perfect environment for mold growth.

ACTIONS AND DECISION-MAKING

After the mold was discovered, several key figures were immediately contacted. The library director was not on campus, so the assistant director took charge. The building coordinator for Neff Annex, the dean of the School of Journalism, and the campus Environmental Health and Safety (EH&S) department were all contacted within the first few hours after the mold was discovered. A preservation staff member was immediately dispatched to the Journalism Library and began assessing damage; likewise, a member of the EH&S staff arrived shortly after. The journalism staff were rightly concerned that the mold could be toxic or otherwise hazardous to the health of those exposed, but the EH&S assessor assured those involved that the mold was not toxic, and should only cause respiratory distress in people with existing allergies. With this in mind, the stack space was immediately closed to users and signage stating the risk to those with allergies was posted at the entrance to the library. Staff working in the infected space were given access to respiratory equipment and training.

After EH&S assured the staff that the mold was not a considerable health risk, the next step concerned stopping the mold growth and cleaning the collection. The author and the preservation staff member began to assess the extent of the mold growth within the collection. Selecting a range in the middle of the space, each book was opened, smelled, and visually examined for mold. These observations were then recorded, and used to extrapolate the extent of damage within the entire collection. After calculating this data, it was estimated that over 60 percent of the collection had visible mold growth, with some areas closer to the ceiling drafts or damp walls in the 90 percent range. The examination of special areas of the collection, such as bound photography magazines, also showed that the dampness was causing warping and solidification of some volumes. Luckily, the photography collection was housed in an area that was only minimally affected. Relative humidity and temperature readings were taken via special monitors, and showed uncontrolled spiking up to 78 percent in both areas (Figure 1).

The preservation staff member, along with the head of the Special Collections, Archives, and Rare Books (SCARaB), recommended that the entire collection be removed from the space and frozen, in order to arrest the growth and force the mold spores to become dormant. This was the course of action outlined in the MU Libraries disaster plan for dealing with the salvage of paper-based items (MU 2005, 4.1), but at the time, the library system did not possess or have access to any freezing facility. In order to follow this course of action, an outside contractor would have to be solicited.

In the meantime, staff had already begun taking measures to combat the mold, using guidelines provided by the websites for SOLINET, Harvard University, and the University of Hawaii. Plastic sheeting was taped over the ceiling holes, in order to stop the humid drafts from enhancing mold growth. Fans were introduced to circulate air, and a closed system HEPA vacuum was borrowed from SCARaB to begin removing the visible mold from the worst areas. Only the outside covers were vacuumed at this time, since no official instructions on how to proceed with clean-up had been handed down.

At this point in the process, disagreements began to arise in how best to deal with cleaning the collection. Both the Preservation Department

FIGURE 1

HOBO Monitor 1

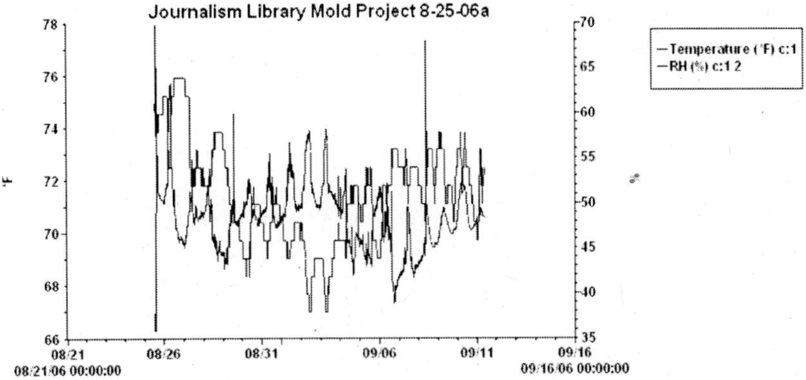

HOBO Monitor 2

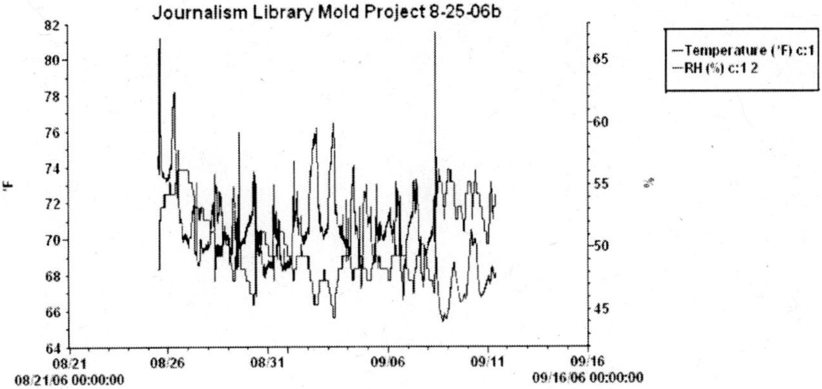

and the head of SCARaB agreed that cleaning should cease and the entire collection should be frozen to stop the mold growth. However, estimates to follow this procedure reached the million dollar range; fiscally, the library could not pursue this option. Alternate recommendations were made for cleaning the items in house using anti-microbial cleansers, and ultimately, these guidelines were followed. The assistant director, aware of the financial repercussions, made the final decision. A list of cleaning and preventative measures was outlined as follows:

1. Continue to vacuum all items in the collection space using the closed system HEPA vacuums.
2. Add additional power outlets to support several dehumidifiers and fans.
3. When possible, wipe the covers of books with isopropyl alcohol or antibacterial spray. Floors, wall, and shelves were to be wiped down with a bleach solution.
4. Books with mold in the inside binding and pages were to be exposed to sunlight to kill spores.
5. Books too badly damaged were to be withdrawn or rebound.
6. Freezing facilities would be purchased, to be available for the worst affected items.

These steps were agreed upon by all parties involved, though everyone was aware that these steps were not the best solution for clean-up or stopping mold growth.

At the same time, the Access Services and Technical Services Departments were developing measures to deal with a suddenly inaccessible popular collection, and to stop the mold contamination from spreading to other parts of the library. The Cataloging Department immediately changed the status of all Journalism items to "temporarily unavailable" and added a message to each item that warned circulation workers to "check for probable mold" during processing. Items transiting back to Journalism were placed into holding rooms by Access Services and lending was restricted. Newspapers and periodicals were still delivered to the infected site, but were kept in a secure location to guard against further infestation. These steps greatly assisted the Journalism Library staff, whose efforts were forced away from user services by the cleaning effort.

The actual cleaning of the collection took place over the next two and a half weeks. Books were removed to carts one shelf at a time. The empty shelf was then wiped down with a bleach solution and dried. Each book was vacuumed separately with a HEPA vacuum. The books were then rubbed with an alcohol moistened micro-fiber cloth to remove any mold spores and placed back on the shelves. Items that were bound in cloth and could not be wiped were exposed to sunlight in order to kill mold spores. Some items, however, were too badly affected to clean. These books, if the pages were still intact, had their covers removed and were placed in tubs for freezing to arrest mold growth before being sent to the bindery. Items too badly damaged even for rebinding were withdrawn. Overall, about 150 books required new bindings; only fifteen sustained sufficient damage to require withdrawal.

By running four dehumidifiers 24 hours a day, the library staff was able to reduce the relative humidity to safe levels. Permanent fans were added to the collection space to encourage air movement, and temperature controls were added specifically for the collection space. The gaps in the ceiling remain, but are blocked by plastic sheeting and are sealed with duct tape to prevent drafts. Even with these additional measures, the collection space must be vigilantly monitored to ensure appropriate environmental conditions.

CONCLUSION: WHAT CAN WE LEARN?

The mold infestation at the MU Journalism Library occurred because of an unusual combination of factors, most of which were weather related and impossible to control. However, the poor renovation of the library space significantly exacerbated the disaster. Repairing the ceiling, added humidity and temperature controls, and fans for air movement have completely arrested mold growth in the collection space. Had these environmental controls been added during the initial renovation phase, it's very likely that the mold growth would not have occurred. This might be a hard lesson for a library with minimal financial resources, but environmental controls should be planned into library space. Prevention will save money, staff time, and materials.

Likewise, the disaster plan of the library system, while comprehensive and structurally sound, was not adequate for this type of disaster. The plan called for action, i.e., the freezing of materials, that was impossible to take at the time. MU Libraries has since purchased freezing facilities to meet the guidelines of the plan, but without this option, the Journalism Library staff were forced to search for other options. All libraries should have a thorough disaster plan in place, with guidelines for dealing with fire, flood, mold, and other issues, but the plan must be realistically within the scope of the library. A non-workable plan does nothing. Planning within the library's means, using readily accessible resources as much as possible, and working with practical solutions are the most important points when crafting a disaster plan. It is also important to establish lines of communication, so each department has full awareness of the situations and can respond appropriately. Designating a primary contact who can make final decisions, a person who has expertise with library disasters, preservation, or other related procedures, is also important, so staff know whom to contact and ask for assistance.

Ultimately, the damages and costs associated with the mold bloom at the MU Journalism Library were kept to a minimum. Quick action by every library department ensured user services continuing as smoothly as possible. Research into alternative cleaning methods proved fruitful and adequate. While communication could have been improved, and the actions taken were not the most ideal course of action, a major disaster was averted and library services were restored relatively quickly. The Journalism Library staff continue preventative measures to protect against further outbreaks, and the collection remains mostly intact as it waits for the new library to open.

REFERENCES

MU Libraries. "Disaster Preparedness, Response and Recovery Manual: Section 4.1 Salvage of Paper Based Materials," University of Missouri-Columbia Libraries. <http://mulibraries.missouri.edu/drp/> (accessed Dec. 4, 2006).

SoliNET Preservation Services. "Invasion of the Giant Mold Spore," Southeastern Library Network. <http://www.solinet.net/preservation/leaflets/leaflets_templ.cfm?doc_id=122> (accessed Nov. 20, 2006).

University of Hawaii at Manoa Preservation Department. "Mold in Hawaii's Libraries," University of Hawaii. <http://libweb.hawaii.edu/libdept/preservation/mold.html> (accessed Nov. 20, 2006).

Weismann Preservation Center. "Library Preservation at Harvard: Guidelines for Managing Mold Contamination," Harvard University Library. <http://preserve.harvard.edu/guidelines/mold.html> (accessed Dec. 6, 2006).

doi:10.1300/J118v25n03_12

Observations of Ground Zero: From the Outside

Sharman Bridges Smith

SUMMARY. Walking into what remains of one of the state's largest public libraries and finding no indication a library was ever there is indescribable. Walls, tables, and shelving gone. No sign of a book, a magazine, a computer, or a video found. Only the terrazzo floors, battered support columns, and a stiff summer breeze blowing unrestricted where large windows once offered spectacular views of the Gulf of Mexico remain. Helping libraries cope with such a disaster is the purpose of this article. The observations and conclusions are those of an outsider looking in, a much safer, but not necessarily wiser, place. doi:10.1300/J118v25n03_13 *[Article copies available for a fee from The Haworth Document Delivery Service: 1-800-HAWORTH. E-mail address: <docdelivery@haworthpress.com> Website: <http://www.HaworthPress.com> © 2006 by The Haworth Press, Inc. All rights reserved.]*

KEYWORDS. Action, communications, community involvement, future, grief, knowledge sharing, preplanning, presence, library, priorities, roles, library

---

Sharman Bridges Smith, MLS, is Executive Director, Mississippi Library Commission, Jackson, MS 39211 (E-mail: sharman@mlc.lib.ms.us).

[Haworth co-indexing entry note]: "Observations of Ground Zero: From the Outside." Smith, Sharman Bridges. Co-published simultaneously in *Public Library Quarterly* (The Haworth Information Press, an imprint of The Haworth Press, Inc.) Vol. 25, No. 3/4, 2006, pp. 151-157; and: *Dealing with Natural Disasters in Libraries* (ed: William Miller and Rita M. Pellen) The Haworth Information Press, an imprint of The Haworth Press, Inc., 2006, pp. 151-157. Single or multiple copies of this article are available for a fee from The Haworth Document Delivery Service [1-800-HAWORTH, 9:00 a.m. - 5:00 p.m. (EST). E-mail address: docdelivery@haworthpress.com].

Available online at http://plq.haworthpress.com

doi:10.1300/J118v25n03_13

INTRODUCTION

On August 29, 2005, Hurricane Katrina destroyed the Mississippi Gulf Coast. While the Coast will recover, things will never be the same. Imagine waking up one morning and everything being gone: no homes, no schools, no churches, no hospitals, no libraries, no grocery stores, no pharmacies, no neighborhoods, no banks, no jobs, no roads, no telephones, no electricity, no water, no food. This is what residents of the Mississippi Gulf Coast faced on August 30, 2005.

Among the devastation, eleven (11) public libraries were heavily damaged or destroyed in Hancock, Harrison, and Jackson counties. The rebuilding of these libraries will take three to five years at best. In addition, 50% of the public library staffs in these Mississippi counties also lost their homes.

Working with these librarians and libraries since the storm has resulted in a number of observations–some obvious, others more subtle, all important to long-term recovery. However, these observations must be prefaced with a disclaimer: "could have," "should have," and "if only" are easier when not living amidst the aftermath of an unprecedented natural disaster. For this reason, this article is written with some apprehension, a lot of "survivor's guilt," and a great admiration for the librarians who are struggling every day to restore public library services.

OBSERVATION: PRIOR RELATIONSHIPS AND PREPLANNING MAKE THE DIFFERENCE IN POST-DISASTER RECOVERY

Planning for a large-scale disaster is not the kind of disaster planning done by most libraries. At one time or another, all of us have attended or sent a staff member to a workshop where a template was distributed and dutifully completed with such information as sources for cleaning supplies, who is responsible for what, and where to get freezer space. This type of information is important, but true disaster planning goes far beyond buying and using cleaning supplies.

In planning, the first question is–What constitutes a disaster? Disasters can be widespread or limited to a given area, such as one building. In any case, when it happens to you, it is a disaster whether it affects anyone else or not. Disasters come in all sizes and from many sources:

- A pipe freezes and bursts in a library, ruining the carpet.
- A car plows through the library's plate glass window.

- A match is dropped in the book drop.
- A tornado sweeps through town, ripping off the library roof.
- A grain silo explodes, dumping grain on the downtown area and the library.
- 40-foot walls of water and 150 miles/hour wind bombard an area for 11 hours.

The kinds of disasters (natural or manmade) vary by location and can be unique to a community or even to a library site. It is important to identify potential disasters in a particular area, whether it is earthquakes, hurricanes, tornadoes, industrial facilities, railroad tracks, heavy snowfall, vandalism, or mechanical failures and plan accordingly.

OBSERVATION: IMAGINE YOUR WORST NIGHTMARE

Most traditional disaster plans are based on a number of positive assumptions. True planning for a disaster should assume only the worst:

- No communications. No phones, no computers, no cell phones, no police radios, no television, no radio. No governing authority: city and county offices destroyed; library board members scattered; no authority in place to respond; no emergency purchase authority.
- No infrastructure. No electricity, no water, no sewer, no roads, no bridges, no police cars, no ambulances, no fire trucks.
- No access to library. Town evacuated, roads impassable, building unstable or toxic, martial law.
- No library staff. Evacuated, unreachable, hospitalized, preoccupied with personal loss, dead.
- No resources. No water, no food, no bank, no money, no people.
- Widespread damage and loss. No homes, no schools, no businesses, no churches.
- No library. Destroyed, commandeered, inaccessible, unsafe.

Even with a clear plan, dealing with such situations is difficult at best. Without a plan, it is almost impossible. Plan for the worst; hope for the best.

OBSERVATION: BE A PLAYER

Being a player is as much about actions taken before a disaster as the ones taken after. In many cases, afterwards is too late. Before a disaster occurs:

- Get actively involved in community organizations, planning, and infrastructure. Become a part of the community's emergency management team.
- Know who community players and gatekeepers are. It is as important these people know you and the roles of the library in the community. Do not wait to be invited; insert the library into the conversation.
- Know whom to turn to for help. Make others aware of how the library can help.
- Share knowledge. Ensure other staff know what you know and have authority to act.
- Be knowledgeable of library insurance coverage even if another entity owns the policy.
- Establish alternative communications links to ensure connections with staff, board members, and governing officials. Collect landline and cell phone numbers, physical addresses, email addresses, and the same information about alternate locations to which people may evacuate. Establish a clearinghouse for messages and information.
- Prearrange business relationships. Establish agreements with salvage, refrigeration, and similar companies to guarantee immediate assistance after a disaster.
- Secure prior emergency authority. Have governing boards or entities delegate, in writing, authority for the library director to act in case of an emergency.
- Set priorities. Not everything can be saved, so what will?

OBSERVATION: GET REAL

Understand the bigger picture and the relationship of the library's issues to the situation. It is critical to acknowledge what is really important.

- Communicate, communicate, communicate! To the extent possible, constant communication with staff, board members, and governing authorities is key.
- Be realistic about larger priorities. Which is more important: saving books or looking for survivors? Finding temporary library facilities or temporary housing for people? Personal disaster or library disaster?
- Chill/grieve. Cut yourself and others some slack. Allow yourself and others time to grieve. Coping with tragedy is as varied as the people who are grieving. Remember that library staff will relive the disaster every day as people come into the library and share their personal stories.
- Access/assess/manage priorities.
    - Where can the library make the most difference at a given point in time? What can the library do to serve the community immediately following a disaster? The library's role may have nothing to do with "traditional" library services and likely will change as recovery moves forward.
    - What should be saved? Remember that most items in public libraries collections are not worth the cost of saving! The obvious exceptions are rare or one-of-a-kind materials such as local history and genealogy materials.

OBSERVATION: TAKE CHARGE

- Practice patience. Understand that nothing happens as quickly as one hopes after a disaster.
- Pick your battles. In light of the larger picture, everything is not equally important. Be part of the solution, not part of the problem.
- Be a presence. Offer solutions. Offer assistance. Make sure the library helps wherever possible even if the tasks have nothing to do with traditional library services. Do what you can to help: maybe the library becomes the county's volunteer coordination center, distributes food stamp cards, offers a cool place with working restrooms, houses the National Guard, or provides the only working Internet connectivity in the county.
- Speak up for the library, but more importantly, speak up for how the library can help.
- Delegate when possible. No one person can do it all.

- Call upon friends and colleagues; most are anxious to help but do not know how.
- Learn to say YES and NO. Be prepared to say what is needed and what is not. Remember, help is only helpful if it does not add to the burden.

OBSERVATION: REESTABLISH A LIBRARY PRESENCE QUICKLY

Library services may need to be redefined temporarily, but the library remains essential to the community, providing a sense of normalcy, representing a sense of place, offering reliable information, and comforting assistance. It is critical to reestablish a library presence quickly for the sake of the community and the future of the library.

OBSERVATION: SEIZE THE OPPORTUNITIES

There are few times in the life of an organization to begin again. Are the library board and staff prepared to take advantage of such an opportunity? Have such conversations taken place before a disaster? Perhaps post-disaster is the time to reassess some library basics and to make changes:

- Different library service offerings, service structure, or service delivery?
- Library facilities rebuilt? Rebuilt in the same locations?
- Changes in the composition of library collections?
- Future needs of community?
- Public expectations of the "new" library?

Planning is never easy. It is particularly hard after a disaster. However, it can also be an opportunity to look forward.

Two of the lessons learned from Hurricane Katrina:

1. *People turn to libraries in time of crisis.*
2. *Library services matter to people and communities.*

OBSERVATION: PLAN FOR THE FUTURE

It is the responsibility of librarians and library boards to make sure that public libraries remain vital parts of communities. Planning is critical to that future, with or without a disaster.

doi:10.1300/J118v25n03_13

Rebuilding a High School Library Collection After Hurricane Katrina

Idella Washington

SUMMARY. After Hurricane Katrina struck, many friends, fellow librarians at all levels, alumni, and others came to help in our rebuilding efforts at Benjamin Franklin High School. I developed an action plan that covers stages in the rebuilding process. It is extremely important that a librarian remain focused and flexible in all rebuilding efforts. The plan, procedures, process, needs assessment, and current educational, library, and technological trends are important to successfully rebuild a library. Librarians faced with any kind of disaster are encouraged to use the steps presented as a guide or modify them in rebuilding other libraries. doi:10.1300/J118v25n03_14

KEYWORDS. Disasters in libraries, rebuilding high school libraries, action plans, library media centers, collection development, technology plans, library administration, collaborative planning, leadership

---

Idella Washington, MLS, is Library Media Specialist, Benjamin Franklin High School, 2001 Leon C. Simon Drive, New Orleans, LA 70122 (E-mail: Idella_washington@benfranklinhighschool.org).

[Haworth co-indexing entry note]: "Rebuilding a High School Library Collection After Hurricane Katrina." Washington, Idella. Co-published simultaneously in *Public Library Quarterly* (The Haworth Information Press, an imprint of The Haworth Press, Inc.) Vol. 25, No. 3/4, 2006, pp. 159-178; and: *Dealing with Natural Disasters in Libraries* (ed: William Miller and Rita M. Pellen) The Haworth Information Press, an imprint of The Haworth Press, Inc., 2006, pp. 159-178. Single or multiple copies of this article are available for a fee from The Haworth Document Delivery Service [1-800-HAWORTH, 9:00 a.m. - 5:00 p.m. (EST). E-mail address: docdelivery@haworthpress.com].

Available online at http://plq.haworthpress.com

doi:10.1300/J118v25n03_14

INTRODUCTION

When disaster occurs to libraries, there appear to be a reawakening of the importance of all libraries in our global society. Hurricanes Katrina, Rita, and Wilma demonstrated the importance of libraries unlike any event in recent history. Libraries have always used volunteers. However, when disaster occurred caused by the hurricanes, volunteers came from across the country to the Gulf Coast area, and are still assisting in recovery efforts. In the event of a disaster, librarians should have a disaster plan and a recovery plan available. This article explains how a high school librarian developed an action plan to rebuild her library.

In early November after returning from evacuating to Arlington, Texas in advance of Hurricane Katrina, I received a telephone call from Ms. Carol Christen, who was the principal of Benjamin Franklin High School before Katrina. The school was one of the public high schools in the New Orleans Public School System. In our conversation, she talked about the status of the school and said that Ben Franklin had applied to become a charter school through the Orleans Parish School Board. She said that once the charter was approved, she wanted me to return as the librarian. I was honored to have been offered the job to continue as librarian at such a unique school with so many gifted, talented, and outstanding students. However, little did I know the extent of the damage to the school library would surpass anything imaginable?

BACKGROUND

Benjamin Franklin High School is located on the University of New Orleans campus. The school's mission is to promote excellence among New Orleans high school students. Our school has been named the top public high school in the State of Louisiana since the implementation of the state school accountability system and was listed in *Newsweek* (May 8, 2006) as one of the public elite schools in the country. It has also been recognized as the top Advanced Placement testing school in the Southern region. Franklin serves a population of students with high academic potential: 99.5% of each graduating class enters college, winning acceptance to the most prestigious universities in the country and collecting significant numbers of merit scholarships. Franklin has maintained its commitment to its mission statement, encouraging students "to reach their maximum potential as productive, creative citizens of a technology-rich global community." To maximize their potential, students may be enrolled in Advanced

Placement courses in their freshman year and are, by the time they are seniors, enrolled in Advanced Placement courses in core academic areas.

Compared to high schools across the nation, Franklin produces an outstanding number of National Merit Finalists and National Achievement Finalists. In 2005, for example, Franklin produced 25 National Merit Semifinalists and 13 National Achievement Finalists. ACT scores hover around a composite average of 27. SAT scores on the verbal half of the test range from 623 to 641 and on the mathematics half of the test from 618 to 631. Additional information is available on the school's website.[1]

STATUS OF DAMAGE

Once charter status was approved, I visited the school to assess the school and library situation. Our school is located on the University of New Orleans campus on the lakefront in New Orleans. The campus is less than a mile from the 17th Street Canal, the canal in which one of the breaches in the levee occurred. My first trip to the campus was quite eerie. As I drove along the Elysian Fields route, I could not believe what I was seeing. It felt like I was in a twilight zone or better yet a war zone after the war. Along the route, there was no sign of life; all I saw was damaged empty houses, dead grass and trees, fallen trees and debris. Everywhere, everything was brown and uninhabited.

When I arrived at Franklin, signs of life were visible on campus. Amidst all of the debris, a large number of volunteers were assisting with clean-up at the school. It was only when I saw the devastation that I realized the impact Hurricane Katrina had and would have for quite some time on our school environment and the community. I walked around and found it quite astonishing to see so many volunteers, parents, faculty, students, alumni and friends of all ages, all assisting in cleanup efforts. The atmosphere was congenial and everyone was excited to see each other again especially at Franklin. I visited the school on many occasions before school opened on January 17, 2006, the 300th birthday of Benjamin Franklin. On my visits, I sorted through books and materials that had to be discarded.

In assessing the situation, I knew that I had quite a challenge ahead. Since I entered the library profession 31+ years ago, I had encountered one other similar challenge but not of this magnitude. The other challenge was replacing and reorganizing a total library collection due to a fire.

The library is located on the first floor of the building, among other key departments. We lost practically everything at the school on the first floor: band instruments, P.E. equipment, gym, art materials, cafeteria, bookroom, library and administrative offices, records, a computer lab and television studio. With the assistance of volunteers, some of the books and a few other materials were salvaged. Volunteers removed wet books and other items and materials from the library to dumpsters. Salvaged books and other materials were brought to classrooms on the second floor as well as to the cafeteria and the bus drop-off area. The first floor of the building is about three feet off the ground. The library had two and a half to three feet of water in it. Below is an example of the molded library workroom which shows the approximate water level height as well as the mold problem. (See Illustration 1.)

Volunteers from Cedarville University as well as parents, teachers, administrators, and others brought books and materials from the library up flights of stairs and put them in two classrooms; other items were randomly put in different classrooms. Students from Union College shelved salvaged books on the 2nd floor foyer before school opened in January. (See Illustrations 2 and 3.)

PROCESS OF RESPONSE

In Michael Gorman's book *Our Singular Strengths: Meditations for Librarians*, he talks about Ranganathan's Five Laws of Library Science as listed below:

- Books are for use.
- Every book its reader.

ILLUSTRATION 1. Post-Katrina Mold Infested Library Workroom

Reprinted with permission.

ILLUSTRATION 2. Union College Students Shelving Books

Reprinted with permission.

ILLUSTRATION 3. Union College Volunteers Signed Their Work

Reprinted with permission.

- Every reader his book.
- Save the time of the reader.
- The library is a growing organism.

At the end of each law, he offered the following resolutions:

- "I will build collections not for vanity but for use."
- "I will add useful items to the collection and make them accessible to all."
- "I will be the connection between the users of my library and the materials they need."
- "I will always bear in mind that the time of library users is precious."
- "I will remember that libraries grow and develop and will plan accordingly."
- "I will apply fundamental principles to guide me through new situations."[3]

It was with the five laws and resolutions in mind that I began to get a firm grip on my new situation. Once the total situation was vivid in my mind, I began to jot down ideas that I would use to lay out my action plan.

Even though we opened school January 17, 2006 (in the midst of all my planning), I am grateful that my principal permitted me to attend the ALA Midwinter Meeting in January 2006. This was an excellent opportunity for me to network with fellow librarians as well as vendors. My attendance had a positive impact on my collection development procedures and ideas for building more partnerships. As fellow librarians saw me, they expressed support and wanted to know what they could do to assist in my rebuilding efforts. Many vendors expressed the same desire as I visited booths and the combined book exhibit looking and checking for books for future purchases.

After assessing the situation, I began next to evaluate the overall school population (faculty, staff, students) and curriculum, space, library media collection, possible services, school environment, the CEO/Principal's administrative decisions and all of the obstacles to services I was likely to encounter. In the end, it was necessary to make adjustment to services as well as decide on the best approach to use to check all salvaged books. I also began to re-read various chapters in *Information Power: Building Partnerships for Learning*, the Louisiana Department of Education's *Guidelines for School Library Media Programs in Louisiana*, and other professional books and journals to assure myself that I was handling the situation in a manner which kept all leadership functions in line with national, state, and local standards and expectations of library media services. I immediately realized that I had to remain focused and flexible.

Ben Franklin's pre-Katrina collection consisted of over 14,000 books and audiovisual materials, 45 researchable periodical titles for a five-year period, plus equipment and technological tools and production equipment that were accessible to students, teachers, and staff. The library was completely automated; the automation system was on the district server, and the online catalog was accessible from any computer in the school. Our students had access to 12 computers in the library and 25 computers in the computer lab which was located on the back wing of the library. Information literacy skills as well as hands-on demonstrations were taught in the computer lab as well as in the library. Students and teachers scheduled classes for special project presentations in the library classroom. We also had a networked printer for student use at no cost to students. Students used library computers for the online catalog, research work, and typing reports.

PLANNING PROCESS

In my planning process, I was able to develop a workable action plan:

1. Organize
   - Re-think library services.
   - Compile various lists.
   - Develop instructions for students, teachers, and volunteers.
2. Approach
   - Complete assessment.
   - Outline approach.
   - Consider options, hindrances, and obstacles.
   - Locate funding sources.
   - Seek and train volunteers.
   - Develop partnerships.
   - Secure acquisitions.
3. Prioritize
   - Determine specific order to carry out needs assessed.
   - Maintain flexibility.
4. Evaluate total process.

5. Pre-assess planning process.
6. Make necessary changes and adjustments.
7. Finalize planning process.

It was extremely difficult to organize all ideas in the midst of so much uncertainty. During the organizational process, I outlined ideas and possible solutions. After evaluating and refining the list many times, I began to decide on a flexible approach. I might add that this planning process was repeated and revised until I felt that I had a firm grip of how to handle every part of the process. Once completed, I basically followed the outline below. The outline is not necessarily in the order followed. In fact, all items listed were integrated in use.

BASIC PROCEDURAL OUTLINE

1. Identify the new school population, population problems: assess needs and best provisional services.
2. Document ideas, changes, exceptions, and problems (save on flash drive and print hard copies).
3. Compile orders (supplies and other needs).
4. Maintain a file of all ideas, compiled lists, donors, volunteers.
5. Seek help (parents, volunteers, peers, friends, grants, organizations, community groups, university and other libraries).
6. Apply for grants.
7. Check and assess salvaged collection.
8. Compare present versus past collection.
9. Check professional literature as well as local, state and national standards.
10. Identify losses.
11. Identify needs based on the curricula, student and staff requests.
12. Collaborate with teachers, administrators, alumni, fellow librarians, universities and local and national organizations as well as community groups.
13. Establish partnerships.
14. Evaluate the acquisition procedures.
15. Evaluate present collection development plan and revise as needed.
16. Revise gifts policy.
17. Incorporate most recent curriculum ideas for 21st Century learning in the library media program.
18. List curriculum units covered; get input about needs from faculty and students.

19. Research unfamiliar problems confronted (mold).
20. Prioritize with flexibility.

The importance of ordering supplies and writing clear, precise instructions for volunteers was given high priority. Since we had volunteers from far and near calling to offer their services to assist in the rebuilding process, instructions were necessary. At the same time, I began to search online for funding from grants.

The salvaged book collection was located on the second floor foyer of the school, an area with high humidity and heat. Volunteers assisted in rebuilding and sorting through salvaged books and materials. They consisted of parents, alumni, friends, faculty, staff, and local members in the community as well as volunteers from across the country from libraries of all levels.

NEW SCHOOL POPULATION AND CURRICULUM

In an effort to understand the school's environment, I needed detailed information about the post-Katrina school's environment and curricula. Pre-Katrina tenrollment was approximately 940 students compared to approximately 540 Post-Katrina enrollments. Most of our students and staff were displaced. The total community and surrounding areas of Greater New Orleans were definitely not the norm. Most basic services were non-existent. Many students traveled long distances to attend school. Our school was on a block schedule from January through May, 2006. This year, we are back on a seven-period schedule.

We continued our collaborative agreement with the University of New Orleans, which permits Ben Franklin students to check books out of its library. However, since the area had no bus service and students were picked up after school, it was difficult for them to take advantage of the opportunity to check books out.

SALVAGED MATERIALS

Since some of the books and materials were saved, it was important to begin to sanitize all salvaged books and equipment. I checked all salvaged books for mold. Volunteers assisted in the following tasks: checking salvaged equipment, videos and books for mold, and putting sanitized books and shelved books in call number order. Volunteers from Lawrenceville School in New Jersey assisted at this critical time when I was in dire need of help. I decided to use a sanitizer recommended for books. Each book was wiped individually, including the edges, inside, and outside. Once the process began, I realized that many of the books were not usable and had to be discarded. Discards were documented. Some of the equipment and materials that had been checked out by teachers on the 2nd and 3rd floor was saved, but most were affected by the humidity and temperature and did not work. Once sanitized, books were shelved by call numbers, which was quite a challenge–due to space. All saved books will be checked for mold again before they are shelved with new books.

NON-SALVAGED MATERIALS

Books were discarded if they had mold or mildew, glued covers with mold, or if they were discolored, wet, or stained. It was evident that we lost books whose covers were glued with brown glue.[4-5] It was also evident that newer books withstood more in heat and humidity than older books. Nearly 2,900 additional books and numerous pieces of equipment and other items were discarded by the end of June. Most of the equipment housed in the library was ruined. Audiovisual materials, unprocessed books, supplies, PHD+, periodicals, and other miscellaneous materials were a total loss. When the sanitizing process was completed, it became apparent that the last three shelves of books had been in water, and that some from the 4th shelf were damaged. Before the process began, I read various pamphlets on handling materials with mold.[6]

All of the furniture, which consisted of wooden tables and chairs, a sofa and chairs, an end table and other library furniture and file cabinets, was lost. All computers were lost. A few of the items from the two top drawers of the file cabinet were copied and saved. Information on the flash drive, including my online circulation program, was saved. The server in the library crashed but the information was on the school system's server and my flash drive was not damaged. Though it was not damaged, I immediately transferred all information to a new flash drive.

GRANTS, VOLUNTEERS, AND OTHER ASSISTANCE

Since the devastation was so extensive, people from across the country called to offer assistance in our recovery efforts. One of the first things that I did when asked to return to work was to check various websites to look for available grants.

The United States Department of Education's web site page had a link to a page called "Hurricane Help for Schools." Schools or libraries were asked for specific needs and a link was set up for companies or other potential donors to sign up to contribute. I submitted information about my school and received numerous calls from businesses and others offering assistance. We accepted only what we could use. Materials that were sent that we could not use were given to schools that could use them.

The American Library Association home page had a Katrina link on its website. It was on this webpage that I found information about the Hurricane Katrina Relief Fund. Cash donations were being accepted to assist in rebuilding libraries in the Gulf Coast Region. ALA also established an Adopt-A-Library Program in which a designated ALA staff member Michael Dowling coordinated and connected disaster libraries with libraries and librarians who wanted to help. I entered the library's name to be adopted. My school library was adopted by three schools through that program. My library also received many in-kind donations as well as many financial gifts from various schools and librarians across the United States as well as support from a few global libraries.

The American Library Association also sponsored a "Communities Build Libraries" program in which libraries in the Gulf Coast Region could sign up for volunteers to visit their school two days during the Annual Conference in June of 2006. I signed up to have volunteers come to my library for two days. During the conference, a quilt made for the students of Benjamin Franklin High School as a gesture of friendship and solidarity by a group of teenagers and librarians at the Sunrise Dan Pearl Library in Sunrise, Florida, was displayed at the Coretta Scott King Award Breakfast. This kind of support was appreciated.

On the ALA website, I found links to other available grants for Gulf Coast libraries. It meant so much to be able to go to one website and find usable links. Words cannot express the gratitude to all who cared so much about our recovery efforts. It was good to know that people from across the country had in their minds to assist libraries and librarians; many are still assisting us. It takes time to recover. Our hearts will be forever touched by the generosity, care, concern, and physical help of so many librarians, alumni, parents, friends, community groups, and churches who helped at a time when needed. As we continue to receive donations and telephone calls, we realize that we are still remembered.

The New Orleans Public Library received warehouses of donated books. They shared those books with other libraries that had been destroyed or damaged. In order to get the books, I had to go to a designated location and sort through boxes of books to see if I could find high school-level books. During the little time that I had, I was able to obtain at least 40 boxes of books that I will add to my collection. These books were processed by librarians who participated in the American Library Association "Libraries Build Communities" Program during their Annual Conference in New Orleans in June 2006.

GRANTS RECEIVED

My library applied for four grants; National Endowment for the Humanities, Beyond Words: Dollar General School Library Relief Fund, Louisiana Library Association Disaster Relief Fund, and the Laura Bush Foundation Gulf Coast School Library Recovery Initiative Grant. We received our first grant, a $10,000 Beyond Words: Dollar General School Library Relief Fund grant. We also received a $10,000 grant from the Louisiana Library Association Disaster Relief Fund, and a $70,000 grant from the Laura Bush Foundation School Library Recovery Initiative. These funds are a good starting point in the rebuilding process. However, more funds are needed.

The "Beyond Words: Dollar General School Library Grant" was the first one that we received. The press conference, which announced the first grants to the Gulf Coast region recipients, was held in June, 2006 at Benjamin Franklin. The school received a lot of news coverage. Those who attended saw the impact of the hurricane.

In addition, we received a "Louisiana Library Association Disaster Relief Grant." This grant was funded from donations from people, associations, organizations, and companies across the country. ALA contributed $100,000 to the LLA Disaster Relief Fund during its Annual Conference in Alexandria in April 2006.

The Laura Bush Grant announcement was presented in late August at Warren Easton High School in New Orleans. Along with this grant came the opportunity of a lifetime. I, along with some other recipients, was not only invited to participate in a roundtable discussion with President Bush and the First Lady Laura Bush but had the opportunity to take a picture with them. Each school was asked to bring a student, the principal, and the librarian. We each received our picture and will always cherish it.

There were many individual schools that made financial contributions to our library and all were appreciated. Lawrenceville School in New Jersey contributed over $13,601 of proceeds from fundraisers. (See Illustration 4.) Lawrenceville set up an account with my vendor and

I was able to make purchases of my choice. In the last fundraiser, they combined financial donations with the Hill School of Pennsylvania to put in the vendor account. Cedarville University Centennial Library Director for Collection Services sent a check for $2,641.05. A group from Cedarville also assisted with cleanup after the storm. (See Illustration 5.)

ILLUSTRATION 4. Lawrenceville School of New Jersey

Reprinted with permission.

ILLUSTRATION 5. Cedarville University of Cedarville, Ohio

Reprinted with permission.

COLLABORATIVE PARTNERSHIP LSU GRADUATE SCHOOL OF LIBRARY AND INFORMATION SCIENCE

According to Hank Rubin in *Collaborative Leadership: Developing Effective Partnerships in Communities and Schools,* "the very structure of public schools and public service nonprofits demands collaborative skills of their leaders. The mandate of public education and nonprofit leaders is to rally resources and communities of support to ensure the achievement of our social missions. Teachers must develop collaborative skills that will enable us to achieve our mission."[7] He also states that "collaborative leaders are interpersonal and inter-institutional relationship managers and that teachers in the 21st century are village builders who will develop partnerships with other public institutions."[8] It is with this thought in mind and the development of collaboration skills mentioned in *Information Power: Building Partnerships for Learning* that I entered into partnership with Dr. Margie Thomas.

Dr. Margie Thomas, Assistant Professor at the LSU Graduate School of Library and Information Science, and I developed a partnership which began in January, 2006. At the 2006 ALA Midwinter Meeting in San Antonio, we discussed and organized our collaborative partnership which continued after Midwinter via e-mail and telephone. Since I had lost three fourths of my collection, we developed a partnership with her Library Science class. As stated by Dr. Thomas,

> Graduate students in library and information science benefit tremendously when they are able to intellectually link information from the classroom to real working libraries. By modifying an existing assignment in a Young Adult Literature class, we were able to allow teams of students to focus on specific areas identified by Mrs. Washington for collection building. The assignment was to identify at least ten of the best, and the most recently published, titles in a particular category. The lists prepared had to include all necessary ordering information, as well as sources of review or recommendation.
>
> The second project included a Service Learning component that allowed three students to complete a portion of the time required for their Field Experience during the summer working with Mrs. Washington to catalog and process donated books and other administrative tasks. Every student learned by doing, they gained tremendously by working with a strong professional, and they can point with pride to what they individually contributed to the rebuilding of the library at Benjamin Franklin High School in New Orleans.[9]

This partnership developed into one in which students were given the opportunity to perform areas in *Information Power*: teaching and learning, leadership, and collaboration when given the opportunity to supervise volunteers. This was a true experience of the overall administrative function in action of teaching and learning, administrative, and information access as well as critical decision making occurring among other experienced professionals in librarianship.[2]

BOOK COLLECTION AND ONLINE CARD CATALOG

Approximately one month after school began, our server crashed. When it came back up, I was able to use my last backup from the previous year to get the circulation program up and running. In March, when I received a computer, I was able to print out all pre-Katrina items that were in the card catalog. This list along with a list of all saved items was later used to provide me with a better estimate and a realistic list of items lost.

Sanitizing and sorting through all saved items was extremely time-consuming, tedious, and untidy. Gloves and masks were worn. The books had undergone extreme heat and humid conditions since the hurricane. The undesirable atmosphere (heat and humidity) contributed to the complexity of the job. This part of the task was not completed until the end of May. We still did not have air in the area where the books were housed. During that period of time, volunteers assisted in this process. The complexity of the situation increased because once sanitized, books had to be put in Dewey order. All discards were documented. Once the sanitizing process was finished, over 3,000 additional books and materials had to be discarded.

In June, we were able to take a manual inventory. We lost our PHD+ in the flood. After calling fellow librarians whose libraries were opened in a nearby parish and my vendor, I was unable to borrow one. Once completed, a print out of current books and losses was printed in call number order. During this process, problems were documented on the print out and later corrected by me and graduate students from LSU School of Library and Information Science. The students also worked on the complete processing of many donated books.

The graduate students had the opportunity to enhance on the job library leadership skills and program administration by assisting in coordinating activities for volunteers at Franklin. This was done when volunteers from the American Library Association Annual Conference in New Orleans in 2006 sponsored a "Libraries Build Community" program, in which volunteers visited Franklin for two days. Students were also shown on local TV and on Sunday morning national news coverage about Gulf Coast library recovery.

LIBRARY SERVICES, JANUARY-DECEMBER 2006

During the first month of school, I established basic library services to students. The orientation process was different. My main function was to concentrate on providing students with information about areas where they could access the best information using available resources. Even though they could check books out from the University of New Orleans campus, our students and most staff members were displaced; some are still displaced.

When we first returned, most computers were not working. The salvaged books were housed on the second floor foyer without air or heat. This meant that books could not be checked out until they were sanitized. Demonstrations of the use of the Louisiana Union Catalog and the Louisiana Library Connection provided them with skills to search all university catalogs as well as additional databases provided free to all public library users in the state of Louisiana. I might add that this is a part of my orientation process every year. Even though our library was not usable, I provided students with brochures, additional passwords to access databases from school and home, UNO library forms, and public library applications. Students were given brochures and additional passwords to databases that they had access to, library forms for UNO and public library applications, and demonstrations on how to access all parts of the library home page.

In the midst of all of the sanitizing, some of our students were able to check out a few books. Since most public libraries in New Orleans were damaged, I recommended that they use the Jefferson Parish Public Library's main branch which offered public library cards to Orleans Parish residents. This service was very useful to our students.

Every effort has been made to add additional resources to the library home page and inform students and faculty when additional links or databases are added. Six additional databases were donated to the library for one year: three from ABC-CLIO and three from H. W. Wilson. We increased the number of databases and will continue to do so in the future. An evaluation of the usage of the online databases will be made at the end of the school year. We also plan to continue to update our automation program.

ACQUISITIONS

Due to the uniqueness of our school environment, we want to continue to purchase quality books and other resources and technological tools that support the curriculum 100%. With the continuous increase in the cost of books, materials, changing technologies, supplementary items, databases, and other critically needed items, we will always need additional funds to assist in rebuilding a quality library collection that will provide all items needed to assist our students to be able to socially function in a global society, continue to become life-long learners and readers, and become more information literate. Acquisitions play a major role in developing a quality library collection. Results from recent research studies continue to support the need and importance of strong quality school library media centers with certified library media specialist, and their impact on student achievement. In 2005, findings from the Illinois Study support the belief that "powerful libraries–and librarians–do, indeed, make powerful learners."[11]

In order to make sure that appropriate materials, books and resources ordered would be the best for our student environment, I reviewed the Louisiana Department of Education *Guidelines for Library Media Programs* and my current Collection Development Policy. In the Department of Education guidelines, the suggested criteria for collection development by category for a collection to be basic, proficient, and exemplary were used.[12] After writing down ideas about what our students needed, I perused current literature and began to assess needs and update my current collection development plan. My main goal was to select the best materials, resources, equipment, and technological tools that would provide students with necessary resources to enhance all areas of the curriculum. I used a variety of print and online selection aids. I organized all additional thoughts into a development plan that would make the process easy to implement.

ACTION PLAN FOR DEVELOPMENT OF A QUALITY COLLECTION

The action plan for collection development for rebuilding followed the high school collection guidelines and resource alignment as recommended in the Guidelines for Library Media Programs in Louisiana Schools,[12] and *Information Power.*[2]

Acquisition Plan:

- Review and revise current collection development policy.
- Ask the department chair in each discipline to submit a form with recommendations for each respective department.
- Request recommendations from individual teachers, administrators, and students.
- Recommendations of resources in all formats (print, non-print, DVD, CD, software, production equipment, technological tools, online databases, and other resources to enhance the curriculum).
- Read reviews of resources in all formats in journals in all disciplines as well as library journals.
- Review current intellectual freedom statements.
- Check all reputable current and back lists of books and materials (such as ALA Best Books lists, etc.).
- Check books with recommended acquisitions of specific titles and reference materials.
- Request delivery of books and other materials in Dewey order.
- Request invoices in Dewey and title order.
- Request MARC records in a specific format or online.
- Provide additional information needed on MARC records, such as source of funding.
- Add additional information when importing MARC records.
- Know your automation system, features and future possibilities.
- Perform an online collection analysis of salvaged books.
- Use online features to prevent duplication of books and other materials.
- Use all features in *Titlewave* and *Sagebrush* to enhance collection development.
- Evaluate and revise technology plan.

We will be updating our automation system so that our students will have remote access and other features to our collection on the web.

Betty J. Morris talks about the role of school library media specialist as change agent. She says that "As a change agent, the school library media specialist must articulate the goals of the school library media program, communicate with all constituencies within the school, develop networks outside the school, and evaluate progress toward set goals."[10] The final product of all plans will be a quality library collection–one that meets the needs of our students, faculty and administrators, and enhances learning. Our collection will provide the best resources, technological tools, online resources, and databases for all our users. Our library services will enhance the overall curricula. Our library program will encourage and promote lifelong readers and learners who will become productive citizens and creators of ideas and information.

The physical rebuilding of the interior and furnishing of the library is incomplete. Perhaps, at the beginning of the year, our students will have full access to the library. At this moment, the inner structure of the library is incomplete and we do not have shelves or furniture. At least 85% of books and media ordered this school year have arrived. We have hundreds of MARC records from manually processed books to be entered into the online catalog. I continue to face many challenges in the rebuilding process. However, I will continue to approach all obstacles with a positive attitude, and remain focused and flexible. The action plan and revised collection development policy will continue to guide my acquisitions. It will bring positive changes in the rebuilding process. Every effort will be made to keep abreast of all changes in the field and to continue to add current quality items to enhance learning for all students.

NOTES

1. Benjamin Franklin High School, <http://benfranklinhighschool.org> (accessed Oct. 2, 2006).

2. American Association of School Librarians Association for Educational Communications and Technology. *Information Power: Building Partnerships for Learning* (Chicago: American Library Association, 1998).

3. Michael Gorman, *Our Singular Strengths: Meditations for Librarians* (Chicago: American Library Association, 1998), 55-69.

4. The Invasion of the Giant Spore. SOLINET Preservation Program, Leaflet Number 5, Sandra Nyberg, November 1987 (Revised July 11, 2003) <http://www.Solinet.net/preservation/leaflets/leaflets_templ.cf?doc_id=122> (accessed Feb. 11, 2006).

5. "CoOl Conservation online," Resource conservation professionals, Preservation Department of Stanford University Libraries, <http://palimpsest.stanford.edu/bytopic/mold/> (accessed Feb. 25, 2006).

6. Ibid.

7. Hank Rubin, *Collaborative Leadership: Developing Effective Partnerships in Communities and Schools* (Thousand Oaks: Jossey-Bass, 2001), 5-6.

8. Ibid. 31.

9. Margie Thomas interview by the author, November 10, 2006.

10. Betty J. Morris, *Administering the School Library Media Center* (Westport: Libraries Unlimited, 2004).

11. Keith Curry Lance, Maurice J. Rodney, and Christine Hamilton-Pennell, *Powerful Libraries Make Powerful Learners*, The Illinois Study Fact Sheet, February 19, 2005, <http://www.islma.org/resources.htm> (accessed Oct. 15, 2006).

12. *Guidelines for Library Media Programs in Louisiana Schools* (Baton Rouge, Louisiana: State Department of Education, 2004), 32-40.

REFERENCES

Carr, Judith F. and Douglas E. Harris. *Succeeding with Standards: Linking Curriculum, Assessment and Action Planning* (Alexandria: VA, Association for Supervision and Curriculum Development, 2001).

Everhart, Nancy. *Evaluating the School Library Media Center* (Englewood, CO: Libraries Unlimited, 1998).

Fitzgerald, Kathleen A. *Program Evaluation: Library Media Services*. Publication developed by the NSSE in collaboration with the Alliance for Curriculum Reform and representatives from the American Association of School Librarians (Schaumburg, IL: National Study of School Evaluation, 1998).

Loertscher, David V., Blanche Woolls and Janice Felker. *Building a School Library Collection Plan: A Beginning Handbook with Internet Assist* (San Jose, CA: Hi Willow Research and Publishing, 1999).

McDonald, Dana and Melvin Bowie, *Information Services for Secondary Schools* (Westport, CT: Greenwood Press, 1998).

Stefl-Mabery, Joette and Jennifer Goodall Powers. Collaborative, "Problem-Based Learning: University and K-12 Partnerships." *Knowledge Quest* vol. 33, no. 4 (2005):14-16.

Woolls, Blanche. *The School Library Media* Manager. 3rd edition. (Englewood, NJ: Libraries Unlimited, 2004).

doi:10.1300/J118v25n03_14

"Nor Any Drop to Drink": New Orleans Libraries in the Aftermath of Hurricane Katrina

Robert E. Skinner

SUMMARY. Hurricane Katrina's winds and flood waters impacted nearly every library in the New Orleans metropolitan area. The author provides a brief commentary on the extent of damage at the academic libraries before describing in detail the destruction and recovery of the library at Xavier University. The article offers a series of questions that librarians in flood-prone areas might ask in order to prepare themselves and their staffs for the possibility of a major flood and its consequences. doi:10.1300/J118v25n03_15 *[Article copies available for a fee from The Haworth Document Delivery Service: 1-800-HAWORTH. E-mail address: <docdelivery@haworthpress.com> Website: <http://www.HaworthPress.com> © 2006 by The Haworth Press, Inc. All rights reserved.]*

KEYWORDS. Disaster assistance, disaster recovery, emergency communication, flood, Hurricane Katrina, mold remediation, New Orleans

---

Robert E. Skinner, MLS, is University Librarian, Xavier University of Louisiana Library, 1 Drexel Drive, New Orleans, LA 70125 (E-mail: rskinner@xula.edu).

[Haworth co-indexing entry note]: "'Nor Any Drop to Drink': New Orleans Libraries in the Aftermath of Hurricane Katrina." Skinner, Robert E. Co-published simultaneously in *Public Library Quarterly* (The Haworth Information Press, an imprint of The Haworth Press, Inc.) Vol. 25, No. 3/4, 2006, pp. 179-187; and: *Dealing with Natural Disasters in Libraries* (ed: William Miller and Rita M. Pellen) The Haworth Information Press, an imprint of The Haworth Press, Inc., 2006, pp. 179-187. Single or multiple copies of this article are available for a fee from The Haworth Document Delivery Service [1-800-HAWORTH, 9:00 a.m. - 5:00 p.m. (EST). E-mail address: docdelivery@haworthpress.com].

Available online at http://plq.haworthpress.com

doi:10.1300/J118v25n03_15

INTRODUCTION

Although most of us who live in South Louisiana are familiar with the threats posed by tropical storms, many had never experienced more than heavy rains and minor-to-moderate street flooding prior to Hurricane Katrina's landfall on August 29, 2005. However, Katrina was a natural disaster that precipitated a larger, man-made disaster, the consequences of which continue to resonate.

New Orleans area libraries were among the hardest hit institutions when levees collapsed at 9:30 a.m. that morning, allowing Lake Pontchartrain to flood into the city. Eight of the New Orleans Public Library's twelve branches were severely damaged by the flood, and due to the layoff of ninety percent of the system's staff, only five are operational at this writing. Two of the five are bookmobiles, and most operational branches are available for only a limited number of hours per week.

Most of the city's public schools, some of which were already in poor physical condition through years of administrative neglect, also flooded, destroying their libraries in the process. A year later, only about one-third of the public schools are operational under various authorities, and some of those do not seem to have books.

The New Orleans Metro area is home to twelve institutions of higher learning, including two medical schools, and nearly all of the libraries experienced some form of damage and/or interruption of service. A brief resume of local experience includes the following:

- **Delgado Community College**: The library, like other campus buildings, took flood waters. Currently the collection is being stored until renovations are completed.
- **LSU Medical Center**: Located Downtown, the Medical School escaped the flood. However, the library's collections were removed by Belfor USA and operations ceased until January 12, 2006. The Dental School, located on a satellite campus, is still closed, including the Dental School Library.
- **Dillard University**: Perhaps the hardest-hit campus in the city, Dillard experienced wind, flood, and fire damage. Every building was affected, and damage is estimated at several hundred million dollars. Dillard's library received two to three inches of flood water on the ground floor and leakage on the second floor. The entire collection has been removed to a storage facility in Texas, and the library is expected to be off line for renovations until mid-2007.

- **New Orleans Baptist Theological Seminary**: The library escaped the flooding that affected many other campus buildings. However, leakage from wind damage resulted in a significant mold infestation.
- **Notre Dame Seminary**: Located between Uptown and Mid-City New Orleans, Notre Dame experienced minor flooding. It became a temporary base of operations for National Guard troops, during which time the seminary's entire operation moved to temporary quarters at St. Joseph Seminary, located on the north shore of Lake Pontchartrain.
- **Nunez Community College**: Located in devastated St. Bernard Parish, all buildings were affected by flood, and the library received flood waters to a depth of eight feet. Cleanup operations are under way, and librarians are offering limited service.
- **Our Lady of Holy Cross**: Situated on the West Bank of the Mississippi River, Holy Cross received no flood damage and little wind damage. However, it became an Occupational Center for National Guardsmen and New York City firefighters, and the library was pressed into service as a dormitory. During cleanup operations, mold was found in the collection. Remediation is being undertaken by library staff members.
- **Southern University at New Orleans**: Located in the Gentilly section of New Orleans, SUNO was also severely affected by flood. University operations, including the library, are being housed in double-wide mobile homes. Most of the library's collection was destroyed by mold, and library renovations have not yet begun.
- **Tulane University**: In a town that is mostly below sea level, Tulane's Howard-Tilton Memorial Library is one of the few buildings with a basement storage area. This area held the university's microform collection, newspapers, an internationally-known music collection, and a government documents collection that dated from the 1880s. The basement flooded, and damages to the collection are between thirty and thirty five million dollars. Some archival collections, located in a satellite building, were also water damaged. Belfor USA has handled recovery operations, and about seventy-five percent of the books and scores in the music collection have been rescued through freeze-drying. Other materials are characterized as a total loss and have been discarded.
- **University of New Orleans**: The campus, located behind levees on the shore of Lake Pontchartrain, escaped the worst of the flooding. However, wind damage to the roof, vandalism by evacuees, and

inexpert work by recovery personnel led to mold infestation of furniture, computer equipment, and parts of the collection.

- **Xavier University**: Every campus building, including the library, experienced flood damage. The library's ground floor received nearly two feet of water, making necessary the entire gutting and reconstruction of that level. Quick action by BMS Catastrophe prevented mold infestation, and the library reopened on January 12, 2006. Campus damage is estimated at nearly $50,000,000.00.

In the months since the storm devastated the city, we've learned that the catastrophe was less the fault of high winds than of shoddy levee construction and improper maintenance. This, in turn, was exacerbated by inadequate levee oversight by a Levee Board composed of political appointees with no civil engineering experience. Libraries, along with the rest of New Orleans, were caught between the proverbial Rock and Hard Place. They couldn't run, nor could they hide. They couldn't swim, but they could sink.

XAVIER UNIVERSITY: DEALING WITH THE SHOCK, FINDING THE PEOPLE

Once the initial shock of the disaster had passed, the first thoughts of Xavier administrators included assessing the damage to the campus and locating faculty. Both of these proved to be significant challenges. Entry into the flooded city proved impossible and would remain so for a couple of weeks by all but water- and airborne rescuers. After floodwaters receded, roadblocks kept many out as the police and military strove to forestall wide-scale looting of vacant properties.

For the four hundred faculty and staff at Xavier, more-or-less instant communication by land-line, phone mail, and campus e-mail accounts had long been taken for granted. However, when the flood hit the campus, Xavier's communication infrastructure vanished.

In these increasingly electronic times, nearly everyone has a cell phone or Blackberry. However, many of us quickly discovered that there is no directory assistance for these devices. Even when contact could be established with another cell phone user, the destruction of most cell-phone relay towers in South Louisiana meant a strong likelihood that the connection would be dropped in mid-call as outlying relay hubs buckled under the added strain.

E-mail contact through Yahoo, GMail, and Hotmail was equally difficult for the same reason. Few campus administrators, including myself, had ever thought to collect the cell phone numbers and external e-mail addresses of more than a handful of colleagues and key employees. Some employees had neither cell phones nor off-campus e-mail accounts.

Interviews I conducted with the leaders of other academic libraries and the New Orleans Public Library made clear that this was a widespread problem among library administrators. It took each library administration weeks to locate faculty and staff with the available technology. Virtually every library dean or director has, since returning to the city, ordered the creation of a database that includes off-campus e-mail addresses, cell phone numbers, and contact information for next-of-kin.

I had my own library webmaster put together a spreadsheet for contact information, then e-mail it as an attachment to everyone's off-campus e-mail account. By keeping those accounts active, every Xavier Library employee will have access to that emergency contact list wherever he or she happens to land, should such a catastrophe reoccur.

I should note here that Xavier University had, after several previous hurricane scares, created an emergency website through which information and instructions could be relayed to employees with web addresses. Although there was no previous effort, institutionally speaking, to collect alternative phone numbers and e-mail accounts, the university webmaster quickly began to gather these from faculty members and administrators with web access. Soon a fairly comprehensive list was accessible. However, no similar list of staff employees was compiled, a fact that was deeply disturbing to those of us with large departments. As is the case in many libraries, paraprofessional staff head up several vital departments in Xavier's Library Resource Center. Not knowing where they were meant that some organizational knowledge was in danger of being lost, particularly after Xavier University declared financial exigency, and laid off nearly all staff employees. This practice was followed by a number of other area academic institutions, as well.

Dealing with the Damage

Located between the Mid-City and Uptown neighborhoods, Xavier University proved to be particularly vulnerable to the flooding resulting from the collapse of the 17th Street and London Avenue Canal levees. The entire campus received flooding that varied in depth from two to eleven feet, affecting every campus building.

The Library Resource Center, opened in 1993, actually received the least flooding, due to the fact that it was built on a slight rise. Only the ground floor of the building received flood waters, and that to a depth of slightly less than two feet. The Reference and Reserve collections were largely unaffected, and only a single computer was lost.

Xavier's institutional leadership was quick to secure the services of Landis Construction, and both men and equipment were on campus soon after flood waters receded. After receiving recommendations from two other institutions for BMS Catastrophe of Fort Worth, Texas, I asked Xavier's executive vice president to get this firm sub-contracted to Landis for the purpose of library recovery. This proved to be no particular difficulty, as BMS was already in New Orleans working for several other clients.

In order to stem the growth of mold throughout the building, BMS gutted the entire first floor, removed damaged furniture and equipment, and piped HEPA-filtered air throughout the building with flexible ductwork. Subsequent to these actions, BMS personnel subjected every surface and volume in the building to multiple dry-wipes in order to insure the retardation of mold spores. This series of actions appears to have been successful. A year and a quarter later, no sign of mold has been found anywhere in the building or on any part of the collection.

When library faculty and staff returned to the campus early in January 2006, electrical service had been returned to the building, and at least some offices had telephone service. Within ten days, all Library offices experienced a resumption of both phone and e-mail service, thereby making it possible for us to return to normal business when classes resumed in mid-January of 2006.

This sounds as though everything came together without much bother, but in fact, the entire campus electrical system was essentially destroyed by the flood. According to Vice-President for Facilities Management Marion Bracy, every building had to be rewired. He related that for weeks thereafter, every time an attempt was made to reconnect Xavier to the city's main power grid, circuits in one campus building or another shorted out, sending electricians back to square one. Hours of tedious inspection and rewiring were necessary to finally get the university off generator power and back onto the grid.

What Are the Lessons?

The flooding of New Orleans was, in many respects, a freak accident that may never happen again. However, it is an object lesson in just how much can go wrong when a large-scale disaster takes place. The entire world bore witness to the frailty of governmental response at such a time. Failure of government agencies to move quickly and efficiently means that libraries, and the institutions they serve, are essentially helpless in the face of disaster.

Another lesson we learned is how little impact individual librarians can make during recovery operations. At Xavier, the campus was sealed off from the outside world, and only trained professionals wearing protective gear were allowed in the recovery zone. This meant that questions about materials, equipment, and collections had to be made long distance, based on sketchy information, or that unilateral actions were taken by recovery personnel without conferring with library staff.

In order to prepare for the potential of flood, individual librarians might begin to asking themselves a series of questions:

1. Are we located in the flood zone of a river or low-lying coastal area?
2. At what level above the street are library entrances located?
3. If we have a basement, how flood-proof is it?
4. If our entrances are at or just above street level, or if we have basement storage areas, what parts of the collection are located in those areas?
5. How can we rearrange collections in order to place more vulnerable things higher in the building?
6. How much of our collection is digital, and should we move into a greater percentage?

Librarians at private institutions should review the insurance coverage for their collections. With the prices of books and serials increasing eight to ten percent per year, it is highly likely that your collections may be underinsured in today's market.

State-supported institutions should investigate state plans or guidelines for dealing with a campus disaster. Take into account that recovery operations may be delayed since contracts for reconstruction and library recovery will likely be put out for bids. Expect that part or all of your collection may be lost to mold before recovery can even begin.

Consider the possibility that the recovery company the state hires may not be all that familiar with up-to-date methods of dealing with mold infestation, the freeze-drying of books, or other issues that may arise regarding damaged library materials. Be prepared to use your staff to supervise, guide, and assist an inexperienced company as it deals with your building and the contents.

At the top of any librarian's agenda should be an investigation into what kind of campus-wide disaster plan is in place. If such a plan exists, where does the library fit into it? Review that portion of the plan, and, if it seems inadequate in any way, strongly suggest changes that will reflect current thinking.

It is probably worthwhile to investigate the companies that deal in library recovery and establish a relationship with one of them ahead of time. Once a company is chosen, schedule a visit from one of their representatives so he or she may survey the library, study the campus disaster plan, make suggestions for safeguarding vulnerable areas, and perhaps assist in the design of a library flood disaster plan.

Dealing with FEMA

So much has been written and said about the poor performance of the Federal Emergency Management Agency that further comment is unnecessary. However, it bears note that no evidence exists that would point to changes within FEMA that would improve its leadership or performance in the face of another large-scale disaster.

It also bears note that, thus far, no institution in New Orleans has seen any money from FEMA, although many have incurred millions of dollars of debt while reconstructing their campuses. No one can say at this point when, if ever, any Federal reimbursement may materialize.

What About Next Time?

In large measure, New Orleans-area library staffs have found that all they can really do after Hurricane Katrina is address symptoms. Dean Mary Lee Sweat at Loyola University, which experienced relatively little physical damage, plans to replace staff desktop computers with laptops. In case of a future evacuation, staff can take their personal machines with them, along with flash drives that will enable them to store and carry files and projects along with them. This equipment will also facilitate a resumption of inter-staff communication during an extended evacuation.

Tulane's Dean Lance Query has installed shut-off valves on all drains entering his building. This should prevent any future flooding of his basement storage area through storm drains.

Xavier University's Marion Bracy plans the installation of campus transformers and generators at least three to four feet higher. Future campus buildings will also be built higher off the ground. Likewise, the university telephone system will be reinstalled on a higher floor. Because campus flooding was so widespread, more sensitive equipment and facilities will not be placed on the ground floor of any building. He also envisions an electrical loop system that would protect the campus electrical system in the event of flood and make it possible to re-electrify the campus sooner once flood waters have receded. He readily admits, however, that it will be quite a while before resources to accomplish this are identified.

Xavier's library staff will make a greater effort to move electronic equipment and some reference resources to upper floors prior to an evacuation. We hope to continue access to digital resources through the University emergency website in the event of a future shutdown of the campus.

CONCLUSION

In many ways, Hurricane Katrina rewrote the books on what librarians should fear and plan for. Every educational institution in New Orleans has experienced a significant drop in enrollment, and all larger institutions have restructured, eliminating a number of graduate programs. These factors have already resulted in diminished revenue, and the loss of both teaching and library faculty and staff. Materials budgets will likely diminish if low enrollments persist.

Each library has also experienced unexpected retirements and resignations from those who chose not to return to New Orleans following the catastrophe. Those who did are, for the most part, uprooted from their homes, friends, and familiar neighborhoods. Some lost family or friends in the flood, or to untimely deaths hastened by grief. Post-traumatic stress does not happen only to soldiers.

A recent census put the population of New Orleans at less than 188,000, down from a pre-storm population of 450,000. Businesses have closed down and replacements have been slow to come on line. Thousands of people try to make homes in trailers meant for vacation travel.

Perhaps the real lesson to be learned from this catastrophe is that librarians recognize their responsibility to their patrons, and will find ways to serve them in spite of everything. Xavier University lost about a third of its library staff. Of the ones who returned, less than a third are living in their own homes. That they returned to an uncertain future with an eagerness to pick up their lives and careers says a great deal about the human spirit and the kind of people who work in libraries everywhere.

doi:10.1300/J118v25n03_15

Lessons Learned from Katrina: What Really Matters in a Disaster

Kay L. Wall

SUMMARY. While life's everyday occurrences in libraries leave us with a disaster situation, nothing could have prepared libraries for the magnitude of Hurricane Katrina. Devastating winds, flooding, and destruction to the area infrastructure required a type of response seldom experienced. By revisiting the responses in minor to catastrophic disasters, there are lessons to be learned. Some lessons serve as a source for prevention for potential disaster situations, rehabilitation in localized disasters, and reinvention when libraries are reduced to rubble. doi:10.1300/J118v25n03_16 *[Article copies available for a fee from The Haworth Document Delivery Service: 1-800-HAWORTH. E-mail address: <docdelivery@haworthpress.com> Website: <http://www.HaworthPress.com> © 2006 by The Haworth Press, Inc. All rights reserved.]*

KEYWORDS. Disasters, Hurricane Katrina, libraries, disaster preparedness, Mississippi

Kay L. Wall, MLS, is Dean of Libraries, Clemson University, Clemson, SC 29634 (E-mail: KWall@Clemson.edu).

Kay L. Wall was University Librarian at the University of Southern Mississippi in Hattiesburg, MS beginning in 2003 and during Hurricane Katrina.

[Haworth co-indexing entry note]: "Lessons Learned from Katrina: What Really Matters in a Disaster." Wall, Kay L. Co-published simultaneously in *Public Library Quarterly* (The Haworth Information Press, an imprint of The Haworth Press, Inc.) Vol. 25, No. 3/4, 2006, pp. 189-198; and: *Dealing with Natural Disasters in Libraries* (ed: William Miller and Rita M. Pellen) The Haworth Information Press, an imprint of The Haworth Press, Inc., 2006, pp. 189-198. Single or multiple copies of this article are available for a fee from The Haworth Document Delivery Service [1-800-HAWORTH, 9:00 a.m. - 5:00 p.m. (EST). E-mail address: docdelivery@haworthpress.com].

Available online at http://plq.haworthpress.com

doi:10.1300/J118v25n03_16

INTRODUCTION

Disasters are nothing new to libraries. Pipes break, fires erupt, HVAC systems break down, weather intervenes, and a host of other possibilities challenge the function of normal library service. However, even being familiar with the everyday interruptions of service due to deterioration of the library's environment, nothing could have prepared me for the impact of Hurricane Katrina on August 29, 2005.

Living sixty miles inland from the Mississippi Gulf Coast, people pay attention to hurricane tracking as a storm develops. Small talk at coffee shops and at work gravitates toward the weather, and the 2005 hurricane season was no different. Reports late in the week preceding Katrina's landfall were cautious as to predicted course and strength, but the weekend's information brought dread to those in its predicted path. First and foremost, personal safety requirements and supplies were on the minds of those who took the warnings seriously. After that, checking on the University Libraries facilities was the final step in "the routine."

Library personnel, managers, and the Disaster Response Action Team (DRAT) had already begun their routine of disconnecting computers, bringing down servers, and protecting materials in or near identified problem areas in the Hattiesburg campus libraries. Placing pieces of plastic sheeting in Special Collections over stacks near windows known to leak, and clearing desktops of papers were accomplished. Since the libraries are part of a larger entity, there was no need for library personnel to be concerned with other considerations such as the removal of exterior furniture. All that was necessary was a brief mental review of the DRAT manual. After many hurricane experiences under our belts, this seemed to be no different.

BACKGROUND

The University of Southern Mississippi is a dual campus, with the central administration located in Hattiesburg. There are two libraries on this campus: the Joseph Anderson Cook Library and the William D. McCain Library and Archives. At the Gulf Coast campus in Ocean Springs, the three year old 54,000 square foot library is located less than 1/10th mile from the Gulf of Mexico. The Gunter Library at the Gulf Coast Research Center in Ocean Springs, MS is located in the Caylor Building, a cinder block building approximately 1/10th mile from the shoreline. The University of Southern Mississippi Libraries has no library facility at the Jackson County teaching site in Gautier, MS on coastal Highway US 90. Library service is provided by the Mississippi Gulf Coast Community College Learning Center next door. University students at the Stennis Space Center teaching site near Picayune, MS use the Maury Naval Oceanographic Library.

Coastal campuses did their normal hurricane preparation. By mid-day Saturday, the storm severity forecasts meant that personal safety and evacuation were the greater concern and there was little time for extra attention to detail at the library. Since the emergency escalated over the weekend, libraries initiated their plans locally without communicating with each other.

Monday, August 29, 2005 began with rain, high winds, and the loss of power in Hattiesburg. Those who evacuated from the storm's path were able to monitor television news, while those without power relied on battery operated communications. By mid-day, the situation deteriorated as Category Two hurricane force sustained winds of 100 miles per hour were recorded in the Hattiesburg area. With trees falling in this heavily forested area, roads were obstructed and dwellings heavily damaged or destroyed. Due to the storm and its severity, widespread damage for coastal and south Mississippi was recorded. Basic services such as electricity, water, telephone, and all other communications were interrupted for days and weeks inland with the hardest hit areas losing services for months. While major inland thoroughfares were accessible within a few days, many roads remained impassable for weeks. Some areas of the city of Hattiesburg received power as early as two days post-Katrina (including the University), while the majority were without electricity for nearly two weeks. The amount of devastation sixty miles inland, not to mention the more extensive devastation on the Gulf Coast, was unexpected and we were unprepared for the aftermath of Katrina.

EXTENT OF DAMAGE

Hattiesburg, MS Campus

Joseph Anderson Cook Library. The window of the Director for the School of Library and Information Science was blown out and her office destroyed. The floor to ceiling window was essentially one wall of her office. Wet carpet from leaking windows was discovered later, as the carpeted areas were underneath filing cabinets. Additionally, there was minor water damage at the expansion joint on the second floor of the library addition, ceiling tiles were on the floor, and the carpet was damp. The expansion joint was in a hallway, so collections were not directly exposed to water. Power and HVAC were restored September 1.

William David McCain Library and Archives. While the facility sustained no significant damage from rain or wind, there were some leaks around windows on the third floor. Wet carpet under filing cabinets was discovered later. The Halon fire suppression system has been damaged and rendered inoperable, according to the University's Safety Officer, and utilities were restored September 1.

Long Beach, MS Campus

USM Gulf Coast Library. Overall, the facility fared well but it did have some minor structural damage, such as cracks in the stucco exterior. The first floor received 12-18 inches of storm water. The roof and all windows remained intact, while there were minor leaks around the windows on the upper floors. The carpet and soft furniture were soaked and there was mud on the tile floor. Items located on the bottom shelves received water damage, but the flood waters did not reach the second shelf. Since there was no power for three weeks, mold and mildew grew rapidly and rampantly. Mold appeared on the walls of the second floor.

While generator service began September 23, climate control was not consistent. Twelve dehumidifiers were delivered September 27 and had to be emptied twice daily to keep them in operation until normal power was restored months later. Due to the extensive damage of the entire campus, where the twenty-foot storm surge washed away the entire first floor of most buildings, the Long Beach campus began offering classes October 10 at the former Garden Park Hospital. Classes are still being held at this campus, with a reoccupation of the Long Beach campus undetermined.

Ocean Springs, MS, Campus at the Gulf Coast Research Laboratory

Gunter Library. The Gunter Library received approximately 18 inches of flood water. Items on the bottom shelves were ruined in both the main part of the library and in the stacks tower. There was more mud on the floors at this facility than in Long Beach, and BMS Catastrophe began working within a week of the storm to clean and salvage materials. While sometimes sporadic, electrical power was restored on September 2. Due to the storm, the HVAC system was not working properly so mold and mildew continued to grow. Three dehumidifiers were dispatched to the site on September 14. There was more initial collection damage at Gunter Library than at the Long Beach Library, but the earlier restoration of power and climate control meant less long-term damage due to mold and mildew. The Gulf Coast Research Lab began offering classes at the Ocean Springs facility on September 19, 2005.

LESSONS LEARNED

After over a decade of library training and education in disaster preparedness, and then experiencing a major disaster, I have learned that different types of disaster require specific types of preparation and response.

Minor Disaster: A minor disaster affects a non-critical part of the library. Examples of this would be the restroom overflowing and the carpet getting wet outside the door. Its effect is minor and has little impact on collections or service. The length of recovery from a minor disaster may last less than a week, and involves no help or expertise aside from that of your existing employees. No specialized library training is needed to rectify the disaster, and correction does not have any significant impact on the library's budget.

Lessons Learned: Minor disasters are an everyday part of a library's existence. Good facility management may avert potential disasters. By responding as needed, accessing cleaning supplies, knowing how to turn off a water valve, and acquiring supplies such as plastic sheeting and duct tape, minor disasters remain minor.

Localized Disaster: A localized disaster involves damage to library materials and collections. Recovery requires some specialized training in the handling of library materials. There is some disruption of services, and some collections may be rendered inaccessible or even damaged beyond repair. The length of total recovery may last for several months, and the library's budget is affected by this type of disaster.

Lessons Learned: Localized disasters happen more than we care to admit. A disaster plan and a trained disaster response team ensure the most effective reaction in a localized disaster. Formalized education through workshops provides an excellent working knowledge for the Team assigned to respond. In addition, there are many print resources to guide the disaster response team's work. A disaster plan provides a strategy, a description of known and potential facility problems, facility infrastructure information, and a plan for collection priorities. While the disaster plan is vital, practice in salvaging wet materials is even more beneficial. This can be accomplished by soaking some discarded books or other library material in muddy water and attempting salvage methods.

Also, assign the task of tracking expenses associated with the disaster recovery to a member of the library staff and communicate this assignment to all staff. While there may not be a source for recovering those costs, unless one gathers the information during the crisis, it is difficult to reconstruct. Only by maintaining this information can disaster costs be totally recovered.

Major Disaster: In a major library disaster, more than 25% of the library materials or library facilities are affected or damaged. Major disasters result in a significant disruption of services, lasting for several months. Efforts to recover from a major disaster require outside assistance, whether from public or privately contracted sources. Entities outside the library may also be affected by this disaster, and basic services such as electricity may be interrupted for the surrounding area. Personal injury and loss affect some staff in a major disaster. Significant financial resources for recovery are needed to return operations to normal, and library collection losses occur. Recovery efforts in a major disaster result in construction and rebuilding. A typical major disaster would be a tornado or a flood.

Catastrophic Disaster: A catastrophic event the magnitude of Hurricane Katrina brings a response totally different from any of the other categories. In a catastrophic disaster, all communications services are rendered inoperable. All infrastructure services such as water and power are unavailable, with widespread damage to the operation of all basic services. Major damage to transportation routes prohibits others from providing immediate support. Damage to all structures is widespread, and resources for recovery not readily available. Primary needs in a catastrophic disaster center around safety for lives. Basic water, food, shelter, transportation, and medical needs are at the forefront of a catastrophic disaster. The restoration of basic services and reintroduction of a supply chain to the affected area take months to achieve. Complete recovery from a catastrophic disaster takes years, with funding needs beyond the ability of local and even state governmental resources.

Lessons Learned: Through the events of Hurricane Katrina, many important lessons were learned (see below). While it is unlikely that catastrophic disasters will affect most of us in the years ahead, many of these lessons may be applied to any type of disaster.

Communication

The most significant lesson learned had to do with the unforeseen loss of communication. In a worst case scenario, assume no regular telephone service, no cell phone, little mass media, and no internet service. After the loss of a communications infrastructure and the evacuation of coastal residents to places around the country, difficulties arose in locating others to see who was alive or dead. Some of the most effective means of communication were unplanned, such as the use of an informal message board hosted externally. To facilitate communication, prepare ahead of time an established communication network. It cannot be elaborate and is subject to many failings. The library should maintain regularly updated, printed emergency phone number and home address lists. In the case of large operations, communication centers should be established that provide information to media outlets on a regular basis. In the future, it has been suggested that additional communication centers be operated by regional library networks or state library to serve as a link between libraries and displaced employees, much like a message board.

Other communication lessons learned included the posting of signs at central gathering places and near emergency facilities to announce communication meetings. Face to face communication by visiting homes is often the most effective means of gathering and disseminating information.

For the leader of your library, inserting yourself into the official communication process between your constituencies and your next higher authority is vital. Get to your local police department to see what your local administration is doing, or to your central administration location. Be a listener and facilitator, and invite yourself to any meetings of the recovery/communications team.

Human Resources

The larger issues of life are more important than the operation of the library. Concentrate on employees and their family needs for shelter. Many lose all possessions in a catastrophic event or have family members who have. The will be continual needs to meet with insurance adjustors and contractors, even as some work patterns return to normal. Continue to put those life issues ahead of the library's needs. At the same time, recognize the need for staff members to verbalize their experiences and help one another through the trauma. This may disrupt what was known as pre-disaster library service. There will be others who need the stability of work, to provide a normalcy in their disrupted lives, once the library is open. Each staff member's needs will be different, and patience will be needed for months and months after the disaster. Recognize that not everyone will be physically or emotionally able to work and that the trauma will affect employees for long into the future.

The physical health of employees who were not injured because of the disaster will most likely suffer due to environmental hazards. Do not let staff do dangerous things, and watch for ongoing hazards such as outbreaks of mold and mildew. Make certain that the environment at the work site is regularly evaluated by a safety professional.

Access

Because of devastation to either the library itself or the surrounding area, be prepared that you may not have access to the site. Be aware that clearance may necessitate updated immunization to gain access. Seek clearance before approaching the disaster site, and exercise good judgment in the brief evaluation time you may be given. Dress in sturdy shoes and old clothing, bring photo identification, and carry a digital camera, flashlights, paper and pens for gathering as much information as possible. Also provide for your own physical needs (food, water) while in the disaster area, and bring supplies for others as well.

If the library can be inhabited, consider its use as a community information center and place to provide support services that other parts of the local, state, and federal government may not have. The good will may provide a higher priority in the pecking order for recovery.

Financial Resources

Assign a staff member to keep track of expenses and be prepared to assign financial decision making to this person. In a catastrophic disaster, normal financial processes may not be possible, so be prepared to make snap spending decisions. Be prepared to be able to justify reasonable expenditures.

Many of the long term recovery needs will be well beyond the library's financial resources. Federal assistance in the form of grants and other government programs have a relatively simple application process, so take advantage of these opportunities. Private foundations may also offer funding to restore services and replace collections. Since there can be multiple opportunities, coordinate efforts by assigning this responsibility to a staff member.

Collections and the "Top Five"

In a catastrophic disaster, only choose the Top Five most important items or discrete collections and forget about everything else. Make certain that all library personnel know the top five. Candidates for the top five are irreplaceable items. Most would be part or all of a special collection and any other unique materials. None of these items should *ever* be stored or even *temporarily* placed on bottom shelves or on the floor! Be careful that only selective items receive special treatment, because the total cost of rehabilitation for replaceable items is prohibitive. Resist the immediate offer from others to quickly freeze dry your collection. If your top five are stabilized then decline the offer, and refer those offering the help to other cultural centers that may need support. The cost of rehabilitation will not be worth it for less valuable items. By realizing that most of our existing collections *are* replaceable, we can do a more effective job concentrating on the top five.

Catastrophic Recovery Services

Commercial services can be very helpful, but be certain any contract is fully understood by all parties before work begins. With limited library manpower, the outside stabilization and recovery services can be essential to minimizing damage. The longer wet carpet and humid conditions are allowed to sit, the greater the damage and the greater the recovery cost. If the library is part of a larger entity, be certain that all parties directing clean-up efforts agree on the scope of the services to be provided.

Volunteers

In a catastrophic disaster, you won't be suffering in isolation. Be prepared that the world will knock at the door, offering assistance as you return to normal operations. Be willing to use volunteers and figure out what they can do. The most important aspect is to say yes to some offers of support and no to others. Develop a response plan, and only accept the support of those who can take care of their own life support needs (food, shelter, etc.) or the burden of caring for others will be an added strain. Not everything volunteers send will be useful, so be prepared to triage books and other library materials to other groups or discard.

If you are a volunteer, don't start sending stuff to the disaster area until asked. Sending books can wait, as there may not even be a library for them, much less shelves in a stable environment to hold them. When entering a disaster bring all supplies that are necessary for your existence and work. Those you plan to help do not need the additional burden of providing for others! There may be no local super store, hardware store, laundromat, restaurant, grocery store, motel, or gas station in the vicinity.

Live Recklessly

Be willing to toss traditional library service out the window, and be prepared to become a community/social services center. Forget about old policies (food, drink, children at the worksite, employee dress) until the world outside stabilizes. Balance reason with wisdom, and be what the community needs you to be first. Forget about fees for services such as making copies. By living recklessly, the library will gain more support and good will needed in the future.

CONCLUSION

Disaster brings opportunities that may never occur again. When given the chance to rebuild, there may be new chances to form multi-type library partnerships or multi-type community service centers for public libraries. Creative configurations may be very attractive to potential donors, foundations, or funding agencies. New partnerships that were not possible before could expand the reach of the library into the community. By living more recklessly, one's possibilities for re-inventing library services are endless.

Some of life's hardest lessons can also be the most beneficial. By learning from past events, we can overcome inertia. By talking to others, by giving disaster response plans attention, and by practicing minor salvage of wet materials, some of the mistakes made by others in the past do not have to be your mistakes. Take threats seriously and be prepared to react when a crisis occurs. Be proactive, reasonable, responsible, and at the same time careful when your disaster occurs: because it will occur.

doi:10.1300/J118v25n03_16

The 2004 and 2005 Gulf Coast Hurricanes: Evolving Roles and Lessons Learned for Public Libraries in Disaster Preparedness and Community Services

Paul T. Jaeger
Lesley A. Langa
Charles R. McClure
John Carlo Bertot

SUMMARY. In the aftermath of the 2004 and 2005 Gulf Coast hurricanes, public libraries played many important roles in their communities, though ensuring access to vital information may have been the most critical service. This article explores the participation of public libraries in the recovery of their communities, the implications of this involvement for the place of public libraries in society, and lessons learned and

---

Paul T. Jaeger, PhD, is Assistant Professor, College of Information Studies, the University of Maryland, College Park, MD 20742 (E-mail: pjaeger@umd.edu).

Lesley A. Langa, MA, is Research Associate, Information Institute, Florida State University, Tallahassee, FL 32306 (E-mail: lesleylanga@gmail.com).

Charles R. McClure, PhD, is Frances Eppes Professor and Director of the Information Institute, Florida State University, Tallahassee, FL 32306 (E-mail: cmcclure@mailer.fsu.edu).

John Carlo Bertot, PhD, is Professor and Associate Director of the Information Institute, Florida State University, Tallahassee, FL 32306 (E-mail: jbertot@fsu.edu).

[Haworth co-indexing entry note]: "The 2004 and 2005 Gulf Coast Hurricanes: Evolving Roles and Lessons Learned for Public Libraries in Disaster Preparedness and Community Services." Jaeger, Paul T. et al. Co-published simultaneously in *Public Library Quarterly* (The Haworth Information Press, an imprint of The Haworth Press, Inc.) Vol. 25, No. 3/4, 2006, pp. 199-214; and: *Dealing with Natural Disasters in Libraries* (ed: William Miller and Rita M. Pellen) The Haworth Information Press, an imprint of The Haworth Press, Inc., 2006, pp. 199-214. Single or multiple copies of this article are available for a fee from The Haworth Document Delivery Service [1-800-HAWORTH, 9:00 a.m. - 5:00 p.m. (EST). E-mail address: docdelivery@haworthpress.com].

Available online at http://plq.haworthpress.com

doi:10.1300/J118v25n03_17

recommendations for public library disaster planning. As this article demonstrates, public libraries successfully provided a range of disaster preparedness and recovery services that were not provided and could not have been provided by other government agencies. doi:10.1300/J118v25n03_17

KEYWORDS. Public libraries, information access, disaster preparedness, internet, computers, disaster recovery

INTRODUCTION

Throughout the course of the 2004 and the 2005 hurricane seasons, the Gulf Coast was battered by a series of tropical storms and hurricanes, many of them major. The shattering impacts of Katrina were the most widely recognized, but a parade of storms–Charley, Dennis, Ivan, Rita, and Wilma, among others–left scars across the states along the Gulf of Mexico. In a little more than a year, Florida alone was hit by eight hurricanes and two tropical storms. These storms affected every part of the Gulf Coast, and a number of communities were hit by multiple storms. Public libraries across the region were playing numerous important roles in their local communities before, during, and after the storms, roles that were not only significant to their patrons, but also served to demonstrate the continuing value of public libraries in service to the community in times of crisis.

The public libraries of Alabama, Florida, Louisiana, Mississippi, and Texas filled a variety of roles in their communities, as they prepared for coming storms, endured the storms, and worked to reassemble their communities after the storms. While the authors recognize that libraries in many other states also responded to the hurricanes by assisting evacuees, providing support to damaged libraries, and filling other needs, this paper focuses attention specifically on the libraries in the Gulf Coast states.

DATA COLLECTION METHODS

The data in this article are drawn from multiple sources. First, the majority of the data for this article was drawn from the 2006 Public Libraries and the Internet study (Bertot, McClure, & Jaeger, 2006). This survey sampled 6,979 public libraries based on three library demographics–metropolitan status (roughly equating to their designation of urban, suburban, or rural libraries), poverty level of their service population (as derived through census data), and state in which they resided. The survey received 4,818 responses, with a response rate of 69%. While completing the survey, 3,887 responding libraries answered the following qualitative question: "In the space below, please identify the single most important impact on the community as a result of the library branch's public access to the Internet." Respondents were able to write as long a response as they desired to the question, and answers ranged from a length of fewer than five words to several paragraphs. This open-ended question provided the most information from the study related to public libraries and hurricanes.

Second, the authors conducted dozens of interviews with public librarians and state library officials along the Gulf Coast regarding their activities, roles, and services in relation to the hurricanes of 2004 and 2005. In all, the authors interviewed dozens of librarians–individually or in small groups–serving a wide range of communities. Further data were provided by some state library organizations that had collected information from libraries affected by the storms.

Through these sources of information, libraries provided detailed accounts of their activities throughout their communities related to the storms. Unless otherwise noted, quotes in this paper are from the qualitative responses to the survey or other information collection activities by the researchers. No alterations have been made to the quotes to correct for minor problems with spelling or grammar, and any such errors in the content are not noted to preserve the original impact of the quotes.

Along with these two streams of data, the authors reviewed professional library literature for further examples about the roles libraries played in the course of the 2004 and 2005 hurricane seasons. Reports from state and regional library journals in the Southeast collected and documented the roles of public libraries after the 2004 and 2005 hurricanes from states along the Gulf Coast from Texas to Florida (Block & Kim, 2006; Choate, 2005; Eberhart, 2005; Fialkoff, 2005; Leboeuf, 2006; LJ News, 2005; McCoy, 2005; Meraz, 2005; O'Connell, 2005; Rogers, 2004). These materials supported the findings of the research reported here and added important examples of libraries contributing to the community.

THE ROLES OF PUBLIC LIBRARIES IN THE EYE OF HURRICANES

A key finding from the 2006 Public Libraries and the Internet study was that the Internet access provided by public libraries in states along the Gulf Coast was of tremendous importance to their communities (Bertot, Jaeger, Langa, & McClure, 2006). The libraries generally described four major roles for the public library Internet access in communities after a hurricane:

1. Locating and interacting with missing, evacuated, and displaced family members and friends;
2. Downloading and filling out Federal Emergency Management Agency (FEMA) forms and insurance claims;
3. Checking for news and updates about conditions in the communities that were evacuated; and
4. Trying to find information about the states of their individual homes or places of work, including checking news sites and satellite maps (Bertot, Jaeger, Langa, & McClure, 2006).

These roles illustrate the importance of public access computing and Internet access in public libraries in the event of natural disaster or other type of crisis. However, the data from the study and the other sources of information also demonstrate that the roles of public libraries in helping their communities deal with the hurricanes extend far beyond the impact of Internet access.

The ways that Gulf Coast public libraries helped their communities cope with the impacts of the 2004 and 2005 hurricanes were wide-ranging and vital. Based on all of the sources of data gathered for this article, these public library roles in their communities before, during, and after the hurricanes include:

1. *Helping communities prepare*–creating and distributing emergency preparedness guides, both printed and web-based; conducting disaster information workshops; and running volunteer coordination programs.
2. *Providing emergency information*–staffing emergency operations centers and public works; answering phone calls; answering email questions; conducting interactive chat services; handling communications in and out of the city; creating community contact centers for community members to re-establish contact; and addressing

inquiries from other parts of the country and from around the world about the conditions in the area or about particular residents.

3. *Giving shelter*–running and staffing shelters for evacuees both in library buildings and in other buildings; providing city employees and relief workers with places to sleep; and housing city command centers for disasters (i.e., police, fire, public works).
4. *Providing physical aid*–cooking and distributing homemade meals; distributing water, ice, meals ready to eat (MREs), tarps, and bug spray; registering people with the "blue roof program"; providing hook-ups to recharge electronics and communication devices; filling water bottles; letting people use library refrigerators for food and medication; and unloading truckloads of relief supplies.
5. *Caring for community members in need*–assisting with the completion of FEMA, insurance, and other paperwork; caring for special needs and elderly evacuees; working as translators for evacuees; running day camps for children when schools were closed and for children of city employees who had to work unusual hours; holding programs in shelters; providing library materials to evacuees in shelters; establishing temporary libraries in shelters; and sending bookmobiles to devastated areas.
6. *Working with relief organizations*–assisting FEMA, Red Cross, and Army Corps of Engineers personnel in their duties; providing meeting spaces for relief and rescue personnel; providing FEMA, Red Cross, National Guard, and Army Corps of Engineers personnel with a place to meet with residents; providing relief personnel a place to use the Internet, email, and telephones; giving temporary library cards to relief workers; and helping FEMA personnel identify local areas that suffered major damage.
7. *Cleaning up the damage after the storms*–securing city buildings before storms; checking structures for damage; cleaning up debris; and restoring damaged government structures.

These efforts occurred across libraries throughout the region, employing their Internet access and many other resources to assist their communities in hurricane recovery in numerous ways. The hurricanes affecting the Gulf Coast in 2004-2005 created important new roles for public libraries and public librarians in communities.

In Florida particularly, the number of hurricanes created significant familiarity with how to help local communities deal with hurricanes. According to Judith Ring, the Florida State Librarian, "there isn't a single county that wasn't affected" by hurricanes in the 2004 hurricane season alone (Rogers, 2004, p. 15). In response to the Public Libraries and the Internet survey, one librarian wrote, "During hurricane season, we have found that hurricane victims used libraries to get in touch with family and friends." Another librarian wrote, "In times of crisis (hurricane aftermath) we were there to provide connectivity to the outside world; reaching out to such entities as FEMA, Insurance companies and loved ones, etc." The tone of the responses from Florida libraries reflected a familiarity with hurricanes not evident in responses from other Gulf Coast states that have had to deal with the 2004 and 2005 hurricane seasons (Bertot, Jaeger, Langa, & McClure, 2006). In fact, after the 2004 hurricane season, one writer already described the public librarians of Florida as being "familiar with securing their buildings and collections, then waiting for storms to pass" (O'Connell, 2005, p. 4).

THROUGH THE STORMS

Below is a selection of the responses from the research conducted by the authors as provided by the libraries. These particular responses were chosen because they more eloquently and succinctly explain the events and the roles of the libraries, but the experiences related in these quotes are typical of other libraries. The origins of the quotes are not provided and any identifying place names were removed because the responding libraries were guaranteed anonymity in their responses.

> We closed for four days due to staff being assigned to Emergency Management to run shelters, distribute food, water, ice, assist FEMA and Red Cross . . . Librarians assisted in running the shelters, cooking and serving meals, and caring for residents.

> Library staff assisted with water and ice distribution. Library staff worked the information phone bank. Library staff assisted with FEMA paperwork and answered calls for information. Library staff worked at 'refuge sites' during two storms.

> We were the City's Command Center, housing police, fire, public works and city manager for three days. They slept on cots and had FEMA-supplied food and water. The library was never without power, so the officials were able to use our phones, TV and computer resources throughout the storm.

> We have a large number of displaced people who are coming to rely upon the library in ways many of them never expected. I've had so many people tell me that they had never been to a library before they had to find someplace to file a FEMA application or insurance claim. Many of these people knew nothing about computers and would have been totally lost without the staff's help.
>
> Library staff have staffed phones at the [Emergency Operations Center] and at the volunteer reception area; handed out tarps; helped register people for the blue roof program with the Army Corp of Engineers and FEMA; helped hand out water/ice/Meals Ready to Eat; provided programming support at a Hurricane Camp for County employees' children and volunteered at regular shelters and the Special Needs Facility.
>
> The Library's parking lot was used to unload tractor-trailer loads of water and meals ready to eat for the distribution center in the county . . . Relief workers were encouraged to use the library to keep in touch with family and friends through email . . . The Library provided a FEMA team with local maps and help in locating areas that potentially suffered major damage from the storm.

As these quotes demonstrate, public libraries helped their communities in numerous ways.

These quotes, however, fail to convey the expansiveness of library efforts. After Katrina and Rita, libraries in Texas were "inundated" with evacuees seeking to use materials, access the Internet, and even seek shelter (Meraz, 2005, p. 90). Librarians in Houston, Texas, created a library in the Houston Astrodome for evacuees to use; the collection included over 10,000 books, 28 computers with Internet access, and 12 computers for children to use (Meraz, 2005). In the Cajundome in Lafayette, Louisiana, the local library created a computer center and branch library for evacuees (LJ News, 2005). One Mississippi library completed 868 FEMA applications for patrons in a single day (Fialkoff, 2005). In Louisiana, some libraries that survived the storms purchased additional computer stations and provided free copying, printing, and faxing to aid evacuees (Choate, 2005). A library in Baton Rouge housed the relief workers who came to the city because hotels were filled with evacuees (Block & Kim, 2006). Across the Gulf Coast, libraries stayed open more hours than normal, worked with more patrons than they had ever seen, and coordinated the recovery of individuals, families, and communities (Block & Kim, 2006; Eberhart, 2005).

Library networks, library organizations, and state libraries also were keenly active in responding to the crises. The State Library and Archives of Florida worked with public libraries throughout the state in coordination efforts to assist libraries in disaster areas, even providing a clearinghouse of information for recovery and relief efforts after each storm <http://dlis.dos.state.fl.us/DisasterRecovery/>. The Southeastern Library Network (SOLINET) responded to Katrina by creating a website <http://www.solinet.net/Disaster> to match goods and services needed at libraries to donors and contributors (Fialkoff, 2005). The Texas Library Association created a program to provide cash and material donations to libraries in Louisiana, Mississippi, and Alabama (Meraz, 2005). At the national level, the American Library Association has been actively involved in recovery efforts and held its 2006 national conference in New Orleans (Kniffel, 2006). These are just examples of the efforts of librarians, libraries, and library organizations.

The single largest contribution to supporting the libraries of the Gulf Coast was announced in June 2006. In a joint effort, the Bill and Melinda Gates Foundation provided $12.2 million, the Bush-Clinton Katrina Fund provided $5 million, and the Institute for Museum and Library Services provided $500,000 to support and rebuild the libraries of the Gulf Coast. These funds will finance the construction of temporary library facilities, the reconstruction of damaged or destroyed libraries, and the addition of computers and other technologies in the rebuilt libraries (Bill & Melinda Gates Foundation, 2006).

THE INTERNET AND THE AFTERMATH

One of the most striking aspects of the information collected for this article was the impact of the ability of libraries to provide Internet access to survivors, evacuees, and relief workers. While a number of the important roles played by libraries, such as providing shelter and distributing food, were also played by other government agencies and relief organizations, the provision of access to computers and the Internet was a wholly unique and immeasurably important role for public libraries. This access proved vital in several different ways, as the Internet was a means to reach necessary insurance and employment information, e-government information and services, email, and news sources, as well as a means to locate missing loved ones.

In a number of communities, the public library was the only place where Internet access was available after the storms passed. "In the aftermath of Hurricanes Katrina and Rita, we were the only source of Internet access to the FEMA and other websites needed by those who needed it." One librarian described the overall situation as, "The Internet was the only means for displaced persons after Katrina to get information they needed. We were flooded with patrons needing Internet access. More and more people come to the public library every day for email, on-line services, job searching, etc."

The use of the Internet to contact FEMA was the most frequently cited service that libraries provided to their communities. FEMA aid forms can only be completed online and require certain levels of software. In some communities, public libraries were the only places where people could get online and access FEMA forms. One librarian reported, "our staff helped customers file over 45,000 FEMA applications [and] insurance claims."

Along with FEMA forms, insurance claims and employment applications frequently had to be completed online as well. These claims and applications were also essential to people who stayed through the storms and to evacuees. As one librarian explained, "our public access computers have been the only source of communicating with insurance carriers, Federal Emergency Management Agency and other sources of aid."

From an emotional standpoint, one of the greatest impacts of the Internet access provided by public libraries related to searches for lost family, friends, and pets. As one librarian stated, "our computers were invaluable in locating a missing family." Searches were conducted by patrons and by librarians helping them to locate evacuees and search for information about those who stayed behind. The Internet access also allowed patrons to have "contact with family members outside of the disaster area," "communicate with family and friends," and "stay in touch with family and friends due to lack of telephone service."

For those who were evacuated or displaced, the Internet access in public libraries provided the only window through which to see what had happened at home. "We had about 100 Katrina refugees using our computers to check on family members and friends who had been evacuated to other places. They also used our computers to file FEMA applications and check on their homes." The Internet access provided by public libraries was also critical to the relief workers who came to some communities, as these relief workers had no other means of communicating except via email with their organizations.

The significance of public access to computers and the Internet in public libraries has never been more pronounced than in the aftermath of these hurricanes. As one librarian poignantly summarized the situation, "For that time–the computers were a Godsend."

NEW PERCEPTIONS OF PUBLIC LIBRARIES

The roles played by public libraries in their communities after the hurricanes seem to have changed the perceptions of libraries in many areas affected by the hurricanes. In no small part, this results from the fact that libraries were able to provide essential information and a range of services to those in need. "Public access to information was the most important service the library provided in the aftermath of Hurricane Katrina."

In some communities affected by the storms, normal public use of libraries seems to have increased since the hurricanes. One librarian wrote, "Since Hurricane Katrina, our computer needs have increased significantly." Another librarian asserted, "The library is becoming better known in our small community as a center for information of all types, especially since Hurricane Katrina." A large portion of this increase in usage is due to an influx of new patrons who had not previously used the library. As a librarian explained, the storms "brought in a new group of library patrons who had never used the library before. This has been especially true since Hurricane Katrina destroyed the MS coast."

Some research has raised questions about the ability of public libraries in various locations to provide vital assistance in a crisis (Harris, Wathen, & Chan, 2005; Matthews, 2005; Will, 2001). However, it is clear that public libraries provided invaluable information and support to communities and individuals affected by the recent seasons of major hurricanes. Not only did public libraries serve a range of important functions in their communities before, during, and after the storms, but also some of the important services they provided have continued to be of great importance to the new users of the library.

The significance of the role of public libraries in coping with the destruction of the 2004 and 2005 hurricane seasons has not been lost on librarians themselves. Librarians have interpreted the response of libraries to these extraordinary events as "breath-taking" (Fialkoff, 2005, p. 8) and "at the heart of providing support to hurricane victims" (Meraz, 2005, p. 90). As providers of support and relief in times of crisis, one commentator asserted, "Librarians may not be glamorous, we don't make a lot of money, but you gotta have us" (Choate, 2005, p. 19).

DISASTER-PREPAREDNESS LESSONS FROM THE GULF COAST

The lessons from the 2004-2005 hurricane seasons for public libraries are myriad. Since forecasters expect the current increase in hurricane activity to continue for years to come (National Oceanic and Atmospheric Administration, 2006), increased discussion, planning, and preparation about the role of public libraries in such situations needs to be expanded. One of best sources for initial planning in disaster preparation and response is dplan <http://www.dplan.org>. This site, prepared by the Northeast Document Conservation Center, provides a broad range of advice and tools to assist in disaster management planning and related activities. Information provided at this site can assist public libraries to develop their own disaster preparedness plan. Such a plan is essential for Gulf Coast states but may also be needed by public libraries in other states for different types of disasters. If they have not done so already, Gulf Coast public libraries must develop a plan for disaster preparedness.

A second lesson is the importance of the state library in assisting public libraries to prepare for a hurricane, supporting their efforts during the crisis, and providing follow-up assistance after the storm. The state library can collect summary data that documents the role of public libraries in the disaster, as was done by the State Library of Florida. Further, it can help inform public libraries and the public as to the role of libraries in such a disaster. The state library can also serve as a clearinghouse for disaster preparedness information, such as making libraries aware of good disaster plans, models for how local libraries can coordinate efforts with local government, and coordinating training related to disaster preparedness. In addition, the state library has the ability to coordinate library services during a hurricane with other state agencies.

Another lesson for local public libraries is to recognize their importance as a community resource in hurricane disaster preparedness. Conversations with Gulf Coast public librarians clearly indicated that the librarians were very surprised by how important the library would be in serving the various roles that have been detailed above in this paper. Recognition of the importance of these roles is a key step to then informing local governing boards, government officials, and residents of what the public library can do to mitigate the effects of a hurricane disaster in a particular community. Disaster preparedness–whether for hurricanes, earthquakes, tsunamis, terrorist attacks, or other types of crises–provides a significant opportunity for the public library to demonstrate its value and importance to the community and its residents.

A further lesson is that public libraries may not be able to rely on immediate outside support after a disaster. In the wake of the problems following the landfall of Katrina, the federal government has made a number of efforts to improve its ability to respond by acquiring better communication and information technologies and by establishing better tracking systems for supplies. However, experts anticipate that federal government responses "will probably still lag in several important areas" (Arnone, 2006, n.p.). A clear lesson from the affected communities during the 2004 and 2005 hurricane seasons is that public libraries should be ready to do as much as possible to help their communities survive and recover from a disaster without relying on immediate outside support from the federal government. As the federal government may or may not be prepared and able to help your community in the aftermath of a disaster, your library needs to be ready to do everything it can independently.

Finally, additional research and study into the roles of public libraries in disaster preparedness need to be done. There is a paucity of such research currently available to public librarians and this lack of information hinders successful disaster preparedness. To a certain degree, the findings presented here have not been widely recognized, have been brought together by piecing different information sources together, and offer a perspective on new and evolving roles for public libraries that requires additional investigation. This paper suggests a host of additional research questions that require attention and investigation.

NEXT STEPS AND RECOMMENDATIONS

The study's findings, and review of additional efforts, point to a number of recommendations to move the public library community forward in preparing to cope with future disasters. These include the need to:

- Involve public libraries directly in disaster planning initiatives. To a large extent, the study showed that public librarians and their libraries often engaged in their relief efforts by accident or through ad hoc efforts, not by design. Given the infrastructure and community presence of public libraries, including them in disaster planning efforts prior to disasters would benefit both the community and disaster relief efforts.

- Clearly define the roles of public libraries in disaster situations within the community. Within the context of community-wide disaster planning, it is important to define specifically the roles that the public library and librarians would play in providing relief and recovery support. This would avoid duplication of efforts, but also enable librarians to engage in training efforts that provide necessary skills, expertise, and practice.
- Provide public librarians with disaster preparedness, relief, and clean-up strategies. Librarians do not typically receive training in hurricane (or other types of disaster) preparedness. The ground zero preparations, plans, resources, and other activities that may occur prior to, during, and after a hurricane (or other disasters) are not well known among public librarians and are unlikely to be taught in schools of library and information science. Thus the state library, regional networks (such as SOLINET in the Southeast), and other organizations may need to rethink the need for such training, the frequency with which it is offered, and the various methods by which such training can be provided.
- Include public librarians and public libraries in disaster training exercises. Community disaster agencies periodically engage in mock or staged training exercises that help prepare emergency service providers for disaster situations. It is important for public libraries and librarians to be included in these exercises so that they too may practice and prepare for relief and clean-up efforts.
- View public library technology infrastructure as critical to disaster recovery and protect it to ensure operability. Public libraries over the years have invested in their technology infrastructure–which includes both wired and wireless Internet access and public access computers (which may include laptops). This infrastructure can be a tremendous asset in times of emergencies, and should be incorporated into community plans.
- Fund public library disaster relief and preparation efforts. Public librarians and libraries by and large offered their services, resources, and assistance without either funding or mandate. They simply stepped up and stepped in. All these efforts, however, were not without a real cost in terms of time, resources, and other tangibles and intangibles. As communities consider public libraries as part of the larger disaster and emergency system, they will need to consider funding libraries for engaging in these efforts.

By including public libraries and librarians in community planning efforts from the onset, those in need will more likely receive efficient and effective service, assistance, and support in the aftermath of disaster.

NEW ROLES AND RESPONSIBILITIES

Public libraries quietly performed an extensive array of services for refugees, evacuees, and survivors of the major storms across the Gulf Coast in 2004 and 2005. "Libraries are, after all, one of the few arms of government that really work" (Fialkoff, 2005, p. 8). This statement was proved countless times in countless communities in the aftermath of Charley, Frances, Jeanne, Ivan, Dennis, Katrina, Rita, Wilma, and the other storms. Katrina may be the most well-known event nationally, but communities from the southern tip of Florida to southern Texas have been severely affected by a number of different storms.

After each storm, people in the affected areas turned to public libraries for many kinds of support and refuge. While all of the forms of aid given by librarians and their libraries proved important in their communities, other government agencies can provide shelter or food or a roof tarp. No other form of government, however, had the public access computers, the Internet access, and the dedicated professionals to turn information into a vital tool for finding the lost, searching for help, requesting aid, and beginning to recover. A librarian in Louisiana explained the immense personal value of libraries by stating, "When you connect someone with his mother and he didn't know whether she was dead or alive, that's important" (Eberhart, 2005, p. 25). Without public libraries, how many thousands of people would still not have been able to fill out FEMA or insurance forms? Apply for a new job? Find displaced loved ones or missing pets? Know what happened in the hometown they had to leave?

For many years, public libraries have been an assumed public good with funding provided to them being primarily viewed as a necessary expense to ensure public access to information and educational materials (Buschman, 2003). However, the changes in government philosophy and policy since the 1980s have altered the political climate for libraries, and many governments now require libraries to provide justification for the funding they receive and the benefits that they offer to their communities (Buschman, 2003; Missingham, 2005). In a political climate where libraries are often asked to justify receiving any amount of public funding, the 2004 and 2005 hurricane seasons have provided, perhaps, one of the best justifications that can be made.

The experiences of public libraries across the Gulf Coast embody many of the community service ideals that libraries strive to fulfill. However, the same experiences merit reflection, as well. Library professionals and researchers need to consider the roles that libraries were able to fill in their communities in times of genuine crisis, what they mean for the education of future librarians, training of current librarians, advocacy for library funding, coordination with other government agencies, relationships with individual patrons, and relationships with entire communities. Being a branch of government that works in times of crisis is an enormous accomplishment, but it is also a serious responsibility. Libraries in other regions of the country, facing other types of disasters, should be ready to perform in the same way the libraries of the Gulf Coast have in the past two years. The issue is how to do so. There is a sizable difference between planning and performing in a crisis.

A key aspect of public libraries across the country becoming better prepared to ensure vital information access and numerous other services during a disaster will be studying Gulf Coast libraries, what they did well in crises, and what might be improved. The 2004 and 2005 hurricane seasons have shown that it is possible for public libraries to respond successfully to a range of crises. A responsibility of all library professionals and researchers is to work toward being ready to serve communities in times of any crisis.

ACKNOWLEDGEMENTS

The 2006 *Public Libraries and the Internet* study was funded by the American Library Association and the Bill & Melinda Gates Foundation. We extend the sincerest gratitude to the thousands of public librarians who made the effort to complete the survey. We also thank all the public librarians and state library officials who participated in the other research related to this paper.

REFERENCES

Arnone, M. (2006, June 5). Storm watch 2006: Ready or not. Federal Computer Week. Available: <http://www.fcw.com>.

Bertot, J. C., Jaeger, P. T., Langa, L. A., & McClure, C. R. (2006). Public access computing and Internet access in public libraries: The role of public libraries in e-government and emergencies. First Monday, 11(9). Available: <http://www.firstmonday.org>.

Bertot, J. C., McClure, C. R., & Jaeger, P. T. (2006). Public libraries and the Internet 2006: Survey results and findings. Tallahassee, FL: Information Use Management and Policy Institute, available at <http://www.ii.fsu.edu/plinternet>.

Bill & Melinda Gates Foundation. (2006). Gulf libraries are building back–and better. Bill & Melinda Gates Foundation. Available: <http://www.gatesfoundation.org>.

Block, M. & Kim, A. (2006). All (librarian) hands on deck. Library Journal, 131(5), 22-25.

Buschman, J. E. (2003). Dismantling the public sphere: Situating and sustaining librarianship in the age of the new public philosophy. Westport, CT: Libraries Unlimited.

Choate, J. (2005). From the president. Louisiana Libraries, Fall 2005, 19.

Eberhart, G. (2005). Katrina's terrible toll: Librarians rally to provide information for a devastated Gulf Coast population. American Libraries, 36(9), 14-25.

Fialkoff, F. (2005). Government that works: A defining moment for public libraries. Library Journal, 130(16), 8.

Harris, R., Wathen, C. N., & Chan, D. (2005). Public library responses to a consumer health inquiry in a public health crisis. Reference & User Services Quarterly, 45(2), 146-154.

Kniffel, L. (2006). You have to see it for yourself. American Libraries, 37(5), 9-13.

Leboeuf, M. C. (2006). Ill winds: Hurricanes and public libraries along the Gulf Coast. Public Libraries, 45(3), 58-63.

"LJ News." (2005). Libraries damaged, librarians respond, after hurricane's fury. Library Journal, 130(15), 16-17.

Matthews, G. (2005). Disaster management: Sharing experience, working across the sector. Journal of Librarianship and Information Science, 37(2), 63-74.

McCoy, M. J. (2005). The forces of nature or dancing with the three ladies. Texas Library Journal, 81(4), 142.

Meraz, G. (2005). In times of trouble. Texas Library Journal, 81(3), 90.

Missingham, R. (2005). Libraries and economic value: A review of recent studies. Performance Measurement and Metrics, 6(3), 142-158.

National Oceanic and Atmospheric Administration. (2006). NOAA: 2006 Atlantic hurricane outlook. Available: <http://www.cpc.ncep.noaa.gov/products/outlooks/hurricane.shtml>.

O'Connell, H. (2005). Weathering the storms: Hurricanes impact Florida libraries. Florida Libraries, 47(1), 4-5.

Rogers, M. (2004). Hurricanes in FL: Floods in wake. Library Journal, 129(17), 15.

Will, B. H. (2001). The public library as community crisis center. Library Journal, 126(20), 75-77.

doi:10.1300/J118v25n03_17

SOLINET's Gulf Coast Libraries Recovery Projects for Public and Academic Libraries

Kate Nevins
Sandra Nyberg

SUMMARY. In 2005, Hurricanes Katrina and Rita damaged many libraries along the Gulf Coast. Due to widespread destruction in communities, library recovery has been slow. With support from the Bill & Melinda Gates Foundation and Andrew W. Mellon Foundation, SOLINET implemented two projects in 2006 to insure that library services remain available in the region. The projects provide temporary public library facilities in communities where libraries were destroyed, distribute grants to restore services at academic libraries and support planning in public libraries, and provide education to facilitate recovery. The projects also address planning for long-term recovery and improved preparedness for future disasters. doi:10.1300/J118v25n03_18 *[Article copies available for a fee from The Haworth Document Delivery Service: 1-800-HAWORTH. E-mail address: <docdelivery@haworthpress.com> Website: <http://www.HaworthPress.com> © 2006 by The Haworth Press, Inc. All rights reserved.]*

Kate Nevins, MLS (E-mail: knevins@solinet.net), is Executive Director, and Sandra Nyberg, MSLS (E-mail: snyberg@solinet.net), is Grants Officer, both at SOLINET, 1438 West Peachtree Street NW, Suite 200, Atlanta, GA 30309-2955.

[Haworth co-indexing entry note]: "SOLINET's Gulf Coast Libraries Recovery Projects for Public and Academic Libraries." Nevins, Kate, and Sandra Nyberg. Co-published simultaneously in *Public Library Quarterly* (The Haworth Information Press, an imprint of The Haworth Press, Inc.) Vol. 25, No. 3/4, 2006, pp. 215-223; and: *Dealing with Natural Disasters in Libraries* (ed: William Miller and Rita M. Pellen) The Haworth Information Press, an imprint of The Haworth Press, Inc., 2006, pp. 215-223. Single or multiple copies of this article are available for a fee from The Haworth Document Delivery Service [1-800-HAWORTH, 9:00 a.m. - 5:00 p.m. (EST). E-mail address: docdelivery@haworthpress.com].

Available online at http://plq.haworthpress.com

doi:10.1300/J118v25n03_18

KEYWORDS. Disaster response, disaster preparedness, Hurricane Katrina, Hurricane Rita, Bill & Melinda Gates Foundation, Andrew W. Mellon Foundation, SOLINET, temporary libraries

This paper describes the early stages of two major grant-funded projects to assist public and academic libraries in recovering from the devastation of Hurricanes Katrina and Rita. Both projects began in summer 2006.

INTRODUCTION

Hurricanes Katrina and Rita damaged and destroyed libraries in a large geographic area along the Gulf Coast of Louisiana, Mississippi, Alabama, and Texas in August and September 2005. Damage spread inland as the storms moved north, especially in Louisiana and Mississippi. Due to many travel restrictions in the hardest hit areas and the general chaos that followed Katrina in particular, it was often weeks before staff members could return to their libraries, by which time the damage caused by the hurricanes had been exacerbated by mold, ongoing exposure to the elements, lack of power, and lack of security. Local resources to assist libraries in response and recovery were very limited in the months following the storms and, in some areas, are still unable to meet the level of need. The hurricanes affected libraries far beyond those actually damaged, as surviving public and academic libraries throughout the region diverted resources to support damaged neighbors. In the months following the hurricanes, the library community stepped forward to provide books and public access computers for displaced populations, library services for students and faculty from campuses that were temporarily closed or inaccessible, office space and communication support for librarians unable to access their own libraries, staff to assist with recovery and salvage, technology support for libraries with damaged communications systems, and storage space for collections that cannot be returned to libraries until repairs are complete.

Damage at individual libraries varied widely, from very minor to complete loss of facilities and collections. Following the storms, the State Library of Louisiana reported 107 public libraries as destroyed or damaged, and the Mississippi Library Commission reported 34 public libraries as either indefinitely or temporarily closed. While some of these libraries have re-opened, 40 in Louisiana and 8 in Mississippi remained closed as of summer 2006, and another 15 were operating with significant service limitations. Thirty-one public libraries were destroyed. Twenty-five academic institutions in the area reported damage to one or more campus libraries. In March of 2006, of 33 damaged academic libraries, 27 were still working to address repair and recovery needs. Three academic libraries were extensively damaged and their original facilities remain closed.

Public libraries typically have more direct control over response and recovery than do academic libraries, where decisions on insurance coverage, building repairs, security, and funding for collection recovery are usually made at the college or university level. Nonetheless, most libraries with minor building damage from Hurricanes Katrina and Rita were repaired and operational by early 2006. Libraries with more extensive building damage were only partially open, not open at all, or operating from temporary or shared facilities in summer 2006. For example, several libraries are able to provide only limited library service through temporary trailers, including the Algiers branch of the New Orleans Public Library, Southern University of New Orleans, the Pass Christian branch of the Harrison County (MS) Public Library, and William Carey University (Hattiesburg, MS).

In addition to addressing building issues, libraries dealt with a range of needs in regard to recovering damaged or lost collections, from identification of what was lost or damaged (and its replacement value) through conservation and treatment of damaged materials, replacement of lost collections, filling gaps in holdings due to lost subscriptions, processing replacements and treated materials, and acquiring or maintaining collections in the face of significantly reduced funding. All area libraries with circulating collections lost books and other material checked-out to patrons at the time of the hurricanes. Unique collections were damaged and lost in both public and academic libraries.

Libraries affected by the hurricanes continue to face potential long-term resource constraints. It may be years before property claims are settled with insurance companies and the Federal Emergency Management Agency (FEMA), in part due to disputes over policy interpretations and limitations. Funds to support current library services remain limited as communities deal with widespread damage and loss, changing community demographics, and rising costs. Expectations for increased funding for library services in the future are low due to decreased tax bases in communities, lower enrollments at colleges and universities, decreased revenue at the state level for all services including public libraries and higher education, and cuts from academic institutions and local and state governments as funds are diverted to support other recovery and infrastructure needs.

As response and recovery proceeded, it became apparent that although Hurricanes Katrina and Rita caused similar types of damage at academic and public libraries, the processes, options, needs, and opportunities for recovery differed. As noted above, some of these differences stem from the systems within which the libraries operate, with academic library decisions driven by their college or university while public libraries are more often on their own in dealing with insurance, FEMA, contractors, response services, etc. Other differences relate to the communities these libraries serve. Academic libraries often did not have to provide services immediately following the storms to their students and faculty because many campuses closed. Public libraries, on the other hand, had immediate and often overwhelming demand for information and services following the storms. In many communities, the remaining public libraries were the only places for residents to access the Internet, necessary for communicating with family and friends, filing insurance claims, and requesting aid from FEMA and state agencies. Although their needs and situations are different, both public and academic libraries face long-term recovery issues. In response to that and with awareness of the distinctions between libraries, the Southeastern Library Network (SOLINET) worked with the Bill & Melinda Gates Foundation and the Andrew W. Mellon Foundation to obtain support for recovering libraries.

THE GULF COAST LIBRARIES PROJECT FOR PUBLIC LIBRARIES

Funded in August 2006 with a $12.2 million grant from the Bill & Melinda Gates Foundation, the goal of the Gulf Coast Libraries Project is to insure that library and public access computer services remain available and accessible to residents and communities in hurricane-impacted areas of Louisiana and Mississippi. Library services are vital to community rebuilding efforts. This project seeks to make certain that citizens have access to the information and services provided by their public libraries and that public libraries remain able to support the immediate and evolving needs of their communities.

Project objectives address immediate, intermediate, and long-term needs at affected public libraries. The immediate need is to provide library services in communities where none exist because damaged or destroyed libraries are closed and not yet rebuilt. This will be done by establishing and providing operational and financial support for 22 temporary library facilities and book/tech mobiles. The intermediate need is to assist public libraries in planning for rebuilding. Assistance will be provided through a multi-day summit on rebuilding for affected library systems and communities, a series of state-based training sessions providing technical planning assistance, and local planning grants for Louisiana parish and Mississippi county systems to assess future needs and implement library building program plans. The long-term need to be addressed through the project is permanent replacement of all lost public access computers in rebuilt libraries. SOLINET is working closely with the State Library of Louisiana, the Mississippi Library Commission, and the affected public library systems to achieve the project's objectives.

Temporary library facilities will enable communities to provide needed services while plans are made for rebuilding. Usable space in permanent structures is at a premium in affected areas, as businesses and organizations seeking to re-establish themselves are quickly leasing the limited amount of space that remains. The transient nature of displaced residents and uncertainty over where residents will rebuild make it more difficult for library systems to commit to long-term leases. A library in a modular trailer is a relatively low-cost solution to providing library services until a permanent facility can be rebuilt. It can be deployed rapidly and, if necessary, be shifted to another site. In addition to trailers, the project will provide book/tech mobiles to serve larger geographic areas with low population densities. Grant funds will support the acquisition of trailers and book/tech mobiles, set-up and delivery, furnishings and collection materials, computers, and telecommunications equipment. The grant also funds selected operating expenses for three years, including staff to provide services at temporary libraries, telecommunications and utilities, and fuel for book/tech mobiles. All temporary libraries and book/tech mobiles will be operational in 2007.

Limited resources are available to support public libraries in planning to rebuild, and many parish and county library systems in Louisiana and Mississippi want advice and expertise to assist in effective planning. Planning initiatives through the Gulf Coast Libraries Project seek to help public libraries insure that new facilities meet current and future needs of their communities. The three components of the project's planning objective–summit, state workshops, and planning grants–will articulate a vision for public libraries in affected areas, provide practical frameworks for planning and developing support, and position the library to move forward quickly when construction funds become available.

To set the vision, "Building Libraries, Building Community: A Summit on the Role of Public Libraries in Recreating Community on the Gulf Coast" was held in November 2006, to engage libraries and their communities in thinking in new ways about library facilities and encourage library systems to "not build back, but build better." As difficult as recovery is, the destruction of older facilities offers an opportunity to redesign libraries and build new facilities capable of delivering services in new ways. Presentations at the Summit explored the roles of a public library in a community, including how it functions as an economic and educational asset, and a social and public space. Asset-based community development principles and tools provided a model for planning that maximizes community involvement in the process. Participants also heard presentations and participated in discussions on trends impacting libraries in Louisiana and Mississippi, including demographic and economic shifts, the plans of state-wide recovery programs, and public perception and use of library services. Participants in the Summit were teams from affected library systems and their communities, including library staff and key leaders critical to rebuilding efforts (library trustees and board members, county/parish planning department heads and architects, and members of library foundations and Friends groups attended). Presentation summaries and videos from the Summit are available on SOLINET's website (http://www.solinet.net). State-specific workshops are scheduled in Louisiana and Mississippi for early 2007, focused on the practical process of planning and building community support, particularly within the context of statewide recovery initiatives.

To support local planning processes, the Gulf Coast Libraries Project will provide planning grants to counties and parishes with destroyed library buildings. Grants will fund a range of activities related to planning for rebuilding, including needs assessment, demographic analysis, building program development, building design and planning processes, fundraising planning and activities, and outreach activities that involve the community in planning and strengthen community support for library rebuilding initiatives. Although the Gulf Coast Libraries Project does not fund actual construction, plans developed through it can position a library to take advantage of other sources of construction money, such as that provided by the Bush-Clinton Katrina Fund through the Americans for Libraries Council.

The final objective of the Gulf Coast Libraries Project is to insure that all public access computers lost to the hurricanes are replaced. While this is dependent in some library systems on rebuilding, it meets a long-standing goal of the Bill & Melinda Gates Foundation, to connect libraries and their patrons to the Internet and insure that they stay connected.

THE ACADEMIC LIBRARIES HURRICANE RECOVERY PROJECT

Funded in July 2006 with an $896,000 grant from the Andrew W. Mellon Foundation, the goals of the Academic Libraries Hurricane Recovery Project are to support short- and long-term recovery needs at affected academic libraries and improve library preparedness for future disasters. Project-supported activities to meet immediate needs include a grant program to assist damaged libraries in re-establishing full service, coordinated and centralized acquisition of discounted and donated products and resources to meet ongoing resource needs, and creation and provision of free information and training on topics that will assist damaged libraries in dealing with ongoing issues related to recovery. Project support for addressing longer-term needs includes consulting for groups of libraries in communities that wish to develop new service models through multi-type cooperative organizations, assessment of the effectiveness of pre-Katrina/Rita preparedness, and organizing a conference to create a regional plan for responding to future disasters.

The Academic Libraries Hurricane Recovery Project was developed following an assessment conducted by SOLINET in early 2006, also funded by the Andrew W. Mellon Foundation. The evaluation process included site visits at damaged academic libraries and review of assessments performed by other groups for public libraries, historical societies, museums, archives and public records offices. A meeting of representatives from all groups identified recovery needs and made recommendations to improve preparedness for and response to future disasters. The report of this assessment, available at SOLINET's website (http://www.solinet.net), highlights the distinctiveness of situations and needs at individual academic libraries.

Although individual needs differed from library to library, all identified the need for funds to assist in acquiring resources to re-establish service levels. Even libraries able to re-open and begin service to returning students and faculty are dealing with ongoing limitations related to the replacement of lost materials and technology, recovery of damaged collections, and staffing. To address the unique needs at affected libraries, a re-grant program through the Academic Libraries Hurricane Recovery Project awarded $365,000 to nineteen academic libraries in November 2006. The grants provided funds to address immediate needs in regard to re-establishing services, including replacement of destroyed equipment, collections, and furnishings; renewal of subscription services for collection resources; mold remediation; and temporary staffing.

While the re-grant program addresses unique service needs at individual libraries, recovering academic libraries will have ongoing needs over the next few years to continue rebuilding collections and services. Combining the needs of all affected academic libraries and devoting central resources to acquiring products, information, and materials for the entire group will enable staff at the libraries to focus directly on the service needs of their patrons. Another short-term need for the affected academic libraries is information and training to address ongoing recovery issues, such as collection valuation, recurring mold, mutual aid agreements, and strategies for rebuilding collections. To address this ongoing and evolving need, the project will support the creation and delivery of free workshops and information resources for affected libraries.

Longer-term, academic libraries, like public, must plan for rebuilding and operating in a changed future. Some communities are interested in exploring new models for providing library service through cooperative programs and shared facilities, combining support for different groups currently served by separate libraries (academic, community college, public, K-12, and private libraries). As communities make plans to rebuild, opportunities exist to re-assess needs within the library environment and beyond. To assist communities interested in investigating cooperative models, the Academic Libraries Hurricane Recovery Project will provide consulting funds for initial planning.

Libraries learned first-hand through the hurricanes about the importance of being prepared, of having up-to-date plans for response and recovery. Whether damaged or not, every library along the Gulf Coast discovered the strengths and weaknesses of training and planning done before the hurricanes. They also have learned what to do differently in the future, what obstacles hampered recovery, and what tools and resources were needed for immediate response. Organizations that support libraries on a state and regional level also learned about limitations in existing disaster response and recovery support programs. To learn from the experience of Katrina and Rita, the Academic Libraries Hurricane Recovery Project includes an assessment of pre-storm preparedness activities and planning for coordinated, effective response to future region-wide disasters. Assessment activities will take place in late 2006 and early 2007, culminating in a report and leading to a region-wide conference on disaster preparedness and response.

Participants in the region-wide conference in 2007 will identify the best strategies for coordinating disaster preparedness, response, and recovery at local, state, and regional levels. Planning for and responding to catastrophic, region-wide disasters is different from dealing with disasters at the individual institution level. In the latter case, local resources are identified to assist in response and recovery, but in a catastrophic situation, such resources are either not available or demand for what resources are available quickly outpaces supply. In catastrophic disasters, many libraries and cultural organizations need help at the same time and assistance efforts may be seriously hampered or delayed due to limited access at damaged sites. During regional disasters, institutions must look outside the local community for assistance at all levels, from response through stabilization to recovery. Planning at the regional level can identify what resources are needed for response, organize supply chains to deliver needed resources, coordinate damage assessment and reporting, coordinate communications, and identify and find support for long-term recovery. A report of recommendations will be produced following the conference.

CONCLUSION

Communities hit by Hurricanes Katrina and Rita will need many years to recover. In some cases, the extent of damage will permanently change the nature of the community. Libraries within these communities will change as the community does; they also may serve as a catalyst for community re-growth. Libraries in particular often provided stability for people following the hurricanes, as a source for information and support and as a symbol of the importance of community.

While the two projects described in this article address the needs of libraries in Louisiana and Mississippi, the results of project activities will benefit all libraries. Resources, tools, and educational programming will help every library be better prepared for the next storm and more able to respond effectively. These resources, along with lessons learned and stories of recovery, will provide models and inspiration to all libraries, some of which will inevitably face disasters of their own in the future. Hurricanes Katrina and Rita have raised awareness of the need to be prepared. The shared experience of the Louisiana and Mississippi libraries will help insure that libraries are prepared in the right way next time. That experience also will help to inform coordination of response in future region-wide disasters and plan for ongoing, long-term support for recovery at affected libraries.

doi:10.1300/J118v25n03_18

Index

